CHEMICAL BONDING AND
MOLECULAR GEOMETRY

■ ■ ■

TOPICS IN INORGANIC CHEMISTRY

A Series of Advanced Textbooks in Inorganic Chemistry

PETER C. FORD, Series Editor

R. Jordan, *Reaction Mechanisms of Inorganic and Organometallic Systems* (1991)
J. Burdett, *Chemical Bonding in Solids* (1995)
R. Gillespie and P. Popelier, *Chemical Bonding and Molecular Geometry: From Lewis to Electron Densities* (2001)

CHEMICAL BONDING AND MOLECULAR GEOMETRY

■ ■ ■

From Lewis to Electron Densities

Ronald J. Gillespie

McMaster University

Paul L. A. Popelier

University of Manchester Institute of Science and Technology

New York   Oxford
OXFORD UNIVERSITY PRESS
2001

Oxford University Press

Oxford   New York
Athens   Auckland   Bangkok   Bogotá   Buenos Aires   Calcutta
Cape Town   Chennai   Dar es Salaam   Delhi   Florence   Hong Kong   Istanbul
Karachi   Kuala Lumpur   Madrid   Melbourne   Mexico City   Mumbai
Nairobi   Paris   São Paulo   Shanghai   Singapore   Taipei   Tokyo   Toronto   Warsaw

and associated companies in
Berlin   Ibadan

Copyright © 2001 by Oxford University Press, Inc.

Published by Oxford University Press, Inc.
198 Madison Avenue, New York, New York, 10016
http://www.oup-usa.org

Oxford is a registered trademark of Oxford University Press

Library of Congress Cataloging-in-Publication Data
Gillespie, Ronald J. (Ronald James)
    Chemical bonding and molecular geometry from Lewis to electron densities / R.J.
Gillespie, P.L.A. Popelier.
        p. cm.—(Topics in inorganic chemistry)
    Includes bibliographical references and index.
    ISBN 0-19-510496-X (ppk), 0-19-510495-1 (cloth)
    1. Chemical bonds—History. 2. Molecules—Models. I. Popelier, P.L.A. II. Title. III.
Series

    QD461.G537 2001
    541.224–dc21   00-025404

Cover Illustration: Representations of the SCl_2 molecule. **Center:** Surfaces of the function $L = -\nabla^2 \rho$ for $L = 0$ au (blue) and $L = 0.60$ au (orange). The $L = 0.60$ surface shows the charge concentrations corresponding to the lone pairs on the sulfur atom and torodial charge concentrations on each chlorine atom (see also Figure 7.5). **Top left:** The Lewis Structure. **Top right:** The VESPR model. **Bottom left:** Contour map of the electron density. **Bottom right:** Contour map of L.

Printing (last digit): 9 8 7 6 5 4 3 2 1

Printed in the United States of America
on acid-free paper

To Lyle Mays

CONTENTS

Preface   xi
Acknowledgments   xiii

Chapter 1   The Chemical Bond: Classical Concepts and Theories   1

1.1   Introduction   1
1.2   Valence   1
1.3   The Periodic Table of the Elements   2
1.4   Structural Formulas   3
1.5   Stereochemistry   5
1.6   The Shell Model   6
1.7   The Ionic Model of the Chemical Bond   8
1.8   The Covalent Bond and Lewis Structures   9
1.9   Polar Bonds and Electronegativity   14
1.10   Polyatomic Anions and Formal Charges   17
1.11   Oxidation Number (Oxidation State)   18
1.12   Donor–Acceptor Bonds   19
1.13   Exceptions to the Octet Rule: Hypervalent and Hypovalent Molecules   20
1.14   Limitations of the Lewis Model   23

Chapter 2   Bond Properties   25

2.1   Introduction   25
2.2   Bond Lengths and Covalent Radii   27
2.3   Multiple Bonds and Bond Order   30
2.4   Ionic Radii   33
2.5   The Lengths of Polar Bonds   37
2.6   Back-Bonding   38
2.7   Bond Dissociation Energies and Bond Enthalpies   39
2.8   Force Constants   42
2.9   Dipole Moments   43

Chapter 3   Some Basic Concepts of Quantum Mechanics   49

    3.1   Introduction   49
    3.2   Light, Quantization, and Probability   50
    3.3   The Early Quantum Model of the Atom   51
    3.4   The Wave Nature of Matter and the Uncertainty Principle   53
    3.5   The Schrödinger equation and the Wave Function   53
    3.6   The Meaning of the Wave Function: Probability and Electron Density   57
    3.7   The Hydrogen Atom and Atomic Orbitals   58
    3.8   Electron Spin   64
    3.9   The Pauli Principle   64
  3.10   Multielectron Atoms and Electron Configurations   69
  3.11   Bonding Models   71
  3.12   Ab Initio Calculations   79
  3.13   Postscript   81

Chapter 4   Molecular Geometry and the VSEPR Model   84

    4.1   Introduction   84
    4.2   The Distribution of Electrons in Valence Shells   85
    4.3   Electron Pair Domains   88
    4.4   Two, Three, Four, and Six Electron Pair Valence Shells   95
    4.5   Multiple Bonds   99
    4.6   Five Electron Pair Valence Shells   106
    4.7   Limitations and Exceptions   110

Chapter 5   Ligand–Ligand Interactions and the Ligand Close-Packing (LCP) Model   113

    5.1   Introduction   113
    5.2   Ligand–Ligand Interactions   116
    5.3   The Ligand Close-Packing (LCP) Model   119
    5.4   Bond Lengths and Coordination Number   122
    5.5   Molecules with Two or More Different Ligands   124
    5.6   Bond Angles in Molecules with Lone Pairs   126
    5.7   Weakly Electronegative Ligands   128
    5.8   Ligand–Ligand Interactions in Molecules of the Elements in Periods 3–6   130
    5.9   Polyatomic Ligands   130
  5.10   Comparison of the LCP and VSEPR Models   132

Chapter 6   The AIM Theory and the Analysis of the Electron Density   134

    6.1   Introduction   134
    6.2   The Hellmann–Feynman Theorem   134
    6.3   Representing the Electron Density   136
    6.4   The Density Difference or Deformation Function   139

6.5   The Electron Density from Experiment   143
6.6   The Topology of the Electron Density   144
6.7   Atomic Properties   153
6.8   Bond Properties   155
6.9   The Diatomic Hydrides of Periods 2 and 3   157
6.10   Summary   161

Chapter 7   The Laplacian of the Electron Density   163

7.1   Introduction   163
7.2   The Laplacian of the Electron Density   164
7.3   The Valence Shell Charge Concentration   165
7.4   The Laplacian and the VSEPR Model   170
7.5   Electron Pair Localization and the Lewis and VSEPR Models   178
7.6   Summary   179

Chapter 8   Molecules of the Elements of Period 2   180

8.1   Introduction   180
8.2   The Relationship Between Bond Properties and the AIM Theory   180
8.3   The Nature of the Bonding in the Fluorides, Chlorides, and Hydrides of
      Li, Be, B, and C   184
8.4   The Geometry of the Molecules of Be, B, and C   197
8.5   Hydroxo and Related Molecules of Be, B, and C   198
8.6   The Nature of the CO and Other Polar Multiple Bonds   202
8.7   Bonding and Geometry of the Molecules of Nitrogen   209
8.8   The Geometry of the Molecules of Oxygen   216
8.9   The Geometry of the Molecules of Fluorine   220

Chapter 9   Molecules of the Elements of Periods 3–6   223

9.1   Introduction   223
9.2   Hypervalence   224
9.3   Bonding in the Fluorides, Chlorides, and Hydrides with an LLP Coordination
      Number Up to Four   231
9.4   Geometry of the Fluorides, Chlorides, and Hydrides with an LLP Coordination
      Number Up to Four   239
9.5   Molecules with an LLP Coordination Number of Five   242
9.6   Molecules with an LLP Coordination Number of Six   250
9.7   Molecules with an LLP Coordination Number of Seven or Higher   251
9.8   Molecules of the Transition Metals   258

Index   259
Formula Index   265

PREFACE

The aim of this book is to provide undergraduate students with an introduction to models and theories of chemical bonding and geometry as applied to the molecules of the main group elements. We hope that it will give the student an understanding of how the concept of the chemical bond has developed from its earliest days, through Lewis's brilliant concept of the electron pair bond up to the present day, and of the relationships between the various models and theories. We place particular emphasis on the valence shell electron pair (VSEPR) and ligand close packing (LCP) models and the analysis of electron density distributions by the atoms in molecules (AIM) theory.

Chapter 1 discusses classical models up to and including Lewis's covalent bond model and Kossell's ionic bond model. It reviews ideas that are generally well known and are an important background for understanding later models and theories. Some of these models, particularly the Lewis model, are still in use today, and to appreciate later developments, their limitations need to be clearly and fully understood.

Chapter 2 discusses the properties of bonds such as bond lengths and bond energies, which provide much of the experimental information on which bonding concepts and explanations of geometry have been mainly based. Again this is a brief summary at a fairly elementary level, serving mainly as a review. No attempt is made to deal with the experimental details of the many different experimental methods used to obtain the information discussed.

In the 1920s it was found that electrons do not behave like macroscopic objects that are governed by Newton's laws of motion; rather, they obey the laws of quantum mechanics. The application of these laws to atoms and molecules gave rise to orbital-based models of chemical bonding. In Chapter 3 we discuss some of the basic ideas of quantum mechanics, particularly the Pauli principle, the Heisenberg uncertainty principle, and the concept of electronic charge distribution, and we give a brief review of orbital-based models and modern ab initio calculations based on them.

Chapter 4 discusses the well-known VSEPR model. Although this model can be regarded as an empirical model that does not directly use quantum mechanical ideas, its physical basis is to be found in the Pauli principle. This dependence on a quantum mechanical concept has not always been clearly understood, so we emphasize this aspect of the model. We have tried to give a rather complete and detailed review of the model, which has been somewhat modified over the years since it was first proposed in 1957.

It has long been recognized that steric interactions between large atoms or groups in a molecule may affect the geometry, and about 40 years ago it was suggested that repulsive interactions between even relatively small atoms attached to a central atom often constitute an important factor in determining molecular geometry. Nevertheless, the importance of ligand–ligand repulsions in determining the geometry of many molecules, which led to the development of the ligand close-packing model, was not clearly established until quite recently. This model, which provides an important and useful complement to the VSPER model, is described in Chapter 5.

In recent years increasingly accurate information on the electron density distribution in a molecule has become available from ab initio calculations and X-ray crystallographic studies. The atoms in molecules (AIM) theory developed by Bader and his coworkers from the 1970s on provides the basis for a method for analyzing the electron density distribution of a molecule to obtain quantitative information about the properties of atoms as they exist in molecules and on the bonds between them. This theory is discussed in Chapters 6 and 7. Unfortunately, AIM has remained until now a rather esoteric mathematical theory whose great relevance to the understanding of bonding and molecular geometry has not been widely appreciated. We give a pictorial and low-level mathematical approach to the theory suitable for undergraduates.

Chapters 8 and 9 are devoted to a discussion of applications of the VSEPR and LCP models, the analysis of electron density distributions to the understanding of the bonding and geometry of molecules of the main group elements, and on the relationship of these models and theories to orbital models. Chapter 8 deals with molecules of the elements of period 2 and Chapter 9 with the molecules of the main group elements of period 3 and beyond.

We welcome comments and suggestions from readers. Please send comments via e-mail to either gillespie@mcmaster.ca or pla@umist.ac.uk. For more information about our research, please visit our web sites—Ronald Gillespie at http://www.chemistry.mcmaster.ca/faculty/gillespie and Paul Popelier at http://www.ch.umist.ac.uk/popelier.htm.

ACKNOWLEDGMENTS

We sincerely thank the following friends, colleagues, and students, who kindly read and commented upon all or parts of the manuscript at various stages in its preparation: Dr. Peter Robinson, Professor Richard Bader, Professor Jack Passmore, Professor Steve Hartman, Dr. George Heard, Dr. Alan Brisdon, Dr. Frank Mair, Ms. Maggie Austen, Mr. Paul Smith, and Mr. Manuel Corral-Valero. We express our gratitude to Professors Wade, Hargittai, and Wiberg, who critically reviewed the entire manuscript and made many useful suggestions for its improvement. We thank Dr. Stephane Noury, Dr. Fernando Martin, Dr. George Heard, and Mr. David Bayles, who prepared many of the figures, and Dr. George Heard, Ms. Fiona Aicken, and Mr. Sean O'Brien for their help in the generation of data. We thank the staff of Oxford University Press for all their assistance and Karen Shapiro, Senior Production Editor, in particular for guiding us so smoothly and competently through the deadlines and intricacies of the production process.

RJG thanks his wife Madge for her encouragement, support, and understanding throughout the whole project, and PLAP thanks his parents for their support.

CHEMICAL BONDING AND

MOLECULAR GEOMETRY

■ ■ ■

1

THE CHEMICAL BOND: CLASSICAL
CONCEPTS AND THEORIES

■ ■ ■

◆ 1.1 Introduction

Whenever two or more atoms are held strongly together to form an aggregate that we call a molecule, we say that there are chemical bonds between them. From the time that the concepts of a molecule and a chemical bond were first developed, chemists have been intrigued by the fundamental question: What is a chemical bond? And by other related questions such as: What forces hold atoms together? Why do atoms combine in certain fixed ratios? and What determines the three-dimensional arrangement of the atoms in a molecule? For many years chemists had no clear answers to these questions. Today, as the result of using a variety of physical techniques, such as X-ray crystallography, electron diffraction, and microwave spectroscopy, we have accumulated detailed information on several hundred thousand molecules. This information, together with the advance in our understanding of the fundamental laws of nature that was provided by the advent of quantum mechanics in the mid-1920s, has led to some reasonably good answers to these fundamental questions, as we discuss in this book. But our understanding is still far from complete and, as new molecules are discovered and synthesized, established ideas often need to be modified. So the nature of the chemical bond is a subject that continues to intrigue chemists. In this chapter we will see how ideas about the chemical bond and molecular geometry developed before the advent of quantum mechanics. Many of these ideas, such as Lewis's electron pair, have been incorporated into the quantum mechanically based theories, and we still use them today.

◆ 1.2 Valence

Observations that compounds have fixed compositions and that therefore their atoms are combined in fixed ratios led to the determination of atomic masses and later to the concept that the atoms of a given element have a characteristic combining power; that is, each atom can form a certain number of bonds called its **valence.** Because a hydrogen atom does not normally combine with more than one other atom, it is given a valence of 1—it is said to be univalent. A chlorine atom, which combines with one hydrogen atom to form the molecule HCl, is also said to have a valence of 1, while an oxygen atom, which forms bonds with two hydrogen atoms to give the molecule H_2O, is said to have a valence of 2, and so on. In other words, the valence of an element is defined as the number of hydrogen or other univalent atoms that it will combine with. For example, the formula of the methane molecule, CH_4, shows that carbon has a valence of 4, and the formula of boron trichloride, BCl_3, shows that boron has a valence of 3. Some elements have several valences. For example, sulfur has a valence of 2 in SCl_2, a valence of 4 in SF_4 and SO_2, and a valence of 6 in SF_6 and SO_3.

◆ 1.3   The Periodic Table of the Elements

The periodic table of the elements proposed by Mendeleev in 1869 was one of the great landmarks in the development of chemistry. Mendeleev showed that when the elements that were known at that time were arranged in order of their atomic weights

Li, Be, B, C, N, O, F, Na, Mg, Al, Si, P, S, Cl, K, Ca, . . . ,

their properties varied in a very regular manner, similar properties recurring at definite intervals. For example, in the series Li, Be, B, C, N, O, F, the properties of these elements change progressively from those of a metal to those of a nonmetal, and the valence increases from 1 for Li up to 4 for carbon and then back to 1 for fluorine, as is illustrated by the formulas of the fluorides of these elements: LiF, BeF_2, BF_3, CF_4, NF_3, OF_2, F_2. The next element, sodium, has properties that closely resemble those of Li and begins a new series (Na, Mg, Al, Si, P, S, Cl) in which each element has properties that closely resemble the corresponding element in the first series, ending with chlorine, which has properties very similar to those of fluorine. Similar series can also be recognized among the heavier elements. Mendeleev took advantage of this regular recurrence of similar properties to arrange the elements in the form of a table, known as the **periodic table** in which elements with similar properties came in the same column of the table (Box 1.1). A modern version of Mendeleev's table is shown in Figure 1.1.

Each vertical column in the table is called a **group,** and each horizontal row is called a **period.** The number of elements in successive periods is

2, 8, 8, 18, 18, 32, (32)

Not all the possible 32 elements in the seventh period are known at the present time. Some of them are very unstable (radioactive), having been synthesized from more stable elements only in recent years, while some remain to be made. The groups numbered 1, 2, and 13–18 are known as the **main groups,** and the 10 groups 3–12, which start in the fourth period, are called the **transition groups.** Some of the groups have special names. For example, the elements in group 1 are known as the alkali metals, those in group 2 as alkaline earth metals, those in group 17 as the halogens, and those in group 18 as the noble gases. Hydrogen appears in group 1 in Figure 1.1 but it is not an alkali metal, although it does become metal-

---

▲ BOX 1.1 ▼
Mendeleev and the Periodic Table

Mendeleev's genius can be appreciated when we remember that only 62 elements were known when he formulated the periodic table. To bring similar elements together in the table, he ignored the atomic masses of a few elements, suggesting that they were incorrect, and he was forced to leave some gaps, which he predicted would be occupied by elements that had not then been discovered, some of whose properties he ventured to predict. It was not until some of these elements were discovered and shown to have properties that agreed well with Mendeleev's predictions that many chemists overcame their initial skepticism about the value of the periodic table. Moreover, the later redetermination of some atomic masses, the discovery of isotopes, and the realization that the order of the elements is based on atomic numbers rather than atomic masses, provided justification for the cases in which Mendeleev ignored the order of atomic masses. Many modifications of Mendeleev's original table have been suggested, but the table in Figure 1.1, which is widely used today, is not very different from that originally proposed by Mendeleev; many additional elements have been incorporated, but without changing the overall structure of the original table. The periodic table not only gave chemists a very useful classification of the elements, but it played a vital role in the elucidation of the structure of atoms and the understanding of valence. Today it still remains a most useful working tool for the chemist.

---

lic at high pressures. Alternatively it could be placed in group 17 because it forms the hydride ion H^- just as the halogens form halide ions such as Cl^-. In fact, hydrogen is a unique element with properties not shared by any other element. In some forms of the periodic table it is not placed in any of the groups. If all the elements in either period 6 or 7 were shown in one row, the table would have an inconvenient shape, so the 14 additional elements in periods 6 and 7 are listed at the bottom of the table. Those in period 6 are the **lanthanide** elements, and those in period 7 are the **actinide** elements.

◆ 1.4   Structural Formulas

Which atoms in a molecule are bonded together was gradually worked out by chemists as they developed the concept of valency. In 1858 Couper represented a bond between the two atoms by a line, as in H—Cl, and this symbol is now universally used. Thus methane may be represented as in Figure 1.2. On the basis of the concept of valence and the compositions of molecules such as ethene (C_2H_4) and sulfur dioxide (SO_2), it became clear that some atoms such as carbon and sulfur can form two or even three bonds to another atom and the symbols $=$ and $\equiv$ were universally adopted as the symbols for double and triple bonds (Figure 1.2). These ideas together with the recognition that carbon atoms in particular could form chains and rings enabled Butlerov in 1864 and Kekulé in 1865 to rationalize what had seemed

The periodic table

Key: atomic number (top), symbol, atomic mass (bottom). Example: 1 / H / 1.008

Period \ Group	1	2	3	4	5	6	7	8	9	10	11	12	13	14	15	16	17	18
1	1 H 1.008																	2 He 4.003
2	3 Li 6.941	4 Be 9.012											5 B 10.81	6 C 12.01	7 N 14.01	8 O 16.00	9 F -19.00	10 Ne 20.18
3	11 Na 22.99	12 Mg 24.30											13 Al 26.98	14 Si 28.09	15 P 30.97	16 S 32.07	17 Cl 35.45	18 Ar 39.95
4	19 K 39.10	20 Ca 40.08	21 Sc 44.96	22 Ti 47.88	23 V 50.94	24 Cr 52.00	25 Mn 54.94	26 Fe 55.85	27 Co 58.93	28 Ni 58.69	29 Cu 63.55	30 Zn 65.39	31 Ga 69.72	32 Ge 72.61	33 As 74.92	34 Se 78.96	35 Br 79.90	36 Kr 83.80
5	37 Rb 85.47	38 Sr 87.62	39 Y 88.91	40 Zr 91.22	41 Nb 92.91	42 Mo 95.94	43 Tc 98.91	44 Ru 101.1	45 Rh 102.9	46 Pd 106.4	47 Ag 107.9	48 Cd 112.4	49 In 114.8	50 Sn 118.7	51 Sb 121.8	52 Te 127.6	53 I 126.9	54 Xe 131.3
6	55 Cs 132.9	56 Ba 137.3	57 La 138.9	72 Hf 178.5	73 Ta 180.9	74 W 183.9	75 Re 186.2	76 Os 190.2	77 Ir 192.2	78 Pt 195.1	79 Au 197.0	80 Hg 200.6	81 Tl 204.4	82 Pb 207.2	83 Bi 209.0	84 Po 209.0	85 At (210)†	86 Rn (222)
7	87 Fr (223)	88 Ra (226.025)	89 Ac (227)	104 Rf (261)	105 Db (262)	106 Sg (263)	107 Bh (262)	108 Hs (265)	109 Mt (266)	110 Uun (Uun)	111 Uuu (Uuu)	112 Uub (Uub)						

Transition Elements: Groups 3–12

Lanthanides (Period 6)

58 Ce 140.1	59 Pr 140.9	60 Nd 144.2	61 Pm 144.9	62 Sm 150.4	63 Eu 152.0	64 Gd 157.2	65 Tb 158.9	66 Dy 162.5	67 Ho 164.9	68 Er 167.3	69 Tm 168.9	70 Yb 173.0	71 Lu 175.0

Actinides (Period 7)

90 Th 232.0	91 Pa 231.0	92 U 238.0	93 Np 237.0	94 Pu (242)	95 Am (243)	96 Cm (247)	97 Bk (247)	98 Cf (251)	99 Es (252)	100 Fm (257)	101 Md (258)	102 No (259)	103 Lr (260)

Figure 1.1 The periodic table.

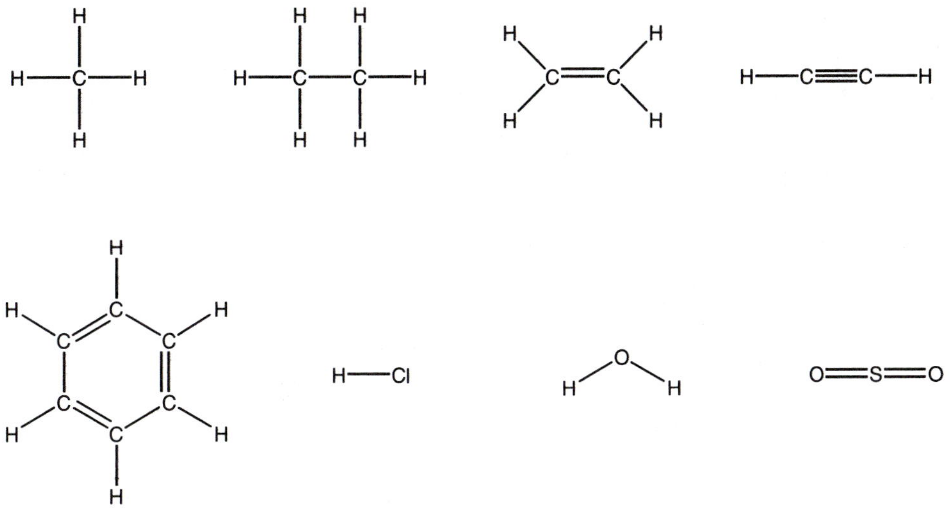

Figure 1.2 Examples of structural formulas.

to be a bewildering variety of formulas for molecules of carbon. For example, Kekulé was able to rationalize the molecular formula C_6H_6 for benzene by the formula in Figure 1.2. The formulas in Figure 1.2, in which the number of lines connected to an atom equal its valence, are examples of what we now call **structural formulas.**

Although the concept of valence worked particularly well for organic molecules and led to a rapid development of organic chemistry, there were many substances, particularly inorganic substances, whose compositions could not be satisfactorily accounted for. For example, some compounds such as $CoCl_3N_6H_{18}$ and K_2SiF_6 had to be represented as "molecular compounds" and given formulas such as $CoCl_3\cdot6NH_3$ and $2KF\cdot SiF_4$ in which two or more molecules whose compositions could be accounted for in terms of the simple concept of valence were supposed to be held together in some unexplained way. The explanation of such compounds had to await the development of a more fundamental understanding of the chemical bond.

◆ 1.5 Stereochemistry

Structural formulas show how the atoms are connected together in a molecule but not how they are they are arranged in space. Indeed, before 1874 chemists had not seriously considered the possibility that the atoms in a molecule might have a definite arrangement in space. In 1874 van't Hoff and le Bel independently proposed an explanation for the existence of **optical isomers**—substances that exist in two forms that have identical physical properties except that a solution of one rotates the plane of polarized light to the left and a solution of the other to the right. At that time around 10 such substances were known, and they were all compounds of carbon in which a carbon atom was bonded to four other different atoms or groups of atoms; that is, they were molecules of the type $CX^1X^2X^3X^4$, where X^1, X^2, X^3, and X^4 are different atoms or groups. Van't Hoff and le Bel proposed that the individual molecules of these substances must therefore exist in left- and right-handed forms that are

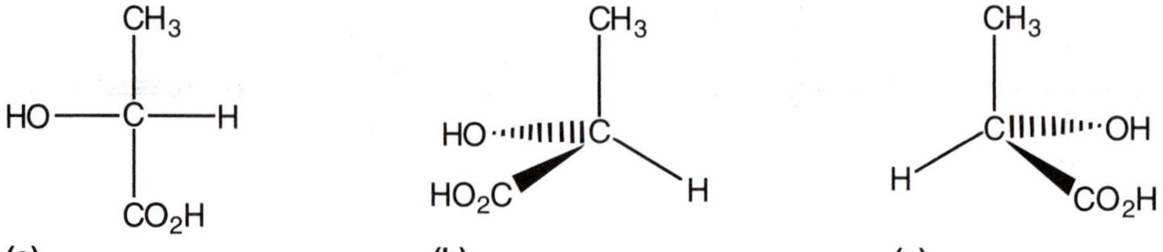

Figure 1.3 Lactic acid. (a) Structural formula. (b) Left- and (c) right-handed enantiomeric forms.

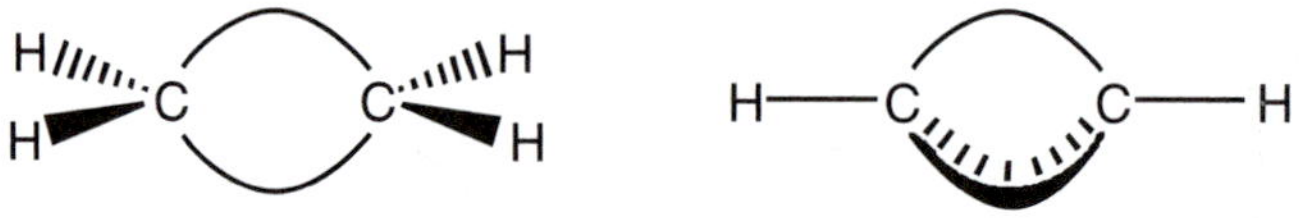

Figure 1.4 Bent bonds in ethene and ethyne.

Figure 1.5 Geometric isomers: the *cis* and *trans* isomers of 1,2-dichloroethene.

mirror images of each other. One form interacts with polarized light to rotate its plane of polarization to the left, while the other rotates it to the right. Molecules of the type $CX^1X^2X^3X^4$ can exist in two mirror image forms only if the four bonds formed by carbon are not in the same plane but are directed toward the corners of a tetrahedron, as shown for lactic acid in Figure 1.3. We now call such molecules **chiral molecules.** Other types of molecule can also be chiral, that is, can exist in right- and left-handed forms.

Double and triple bonds between carbon atoms were then represented by curved lines between the two atoms, to maintain the tetrahedral angle at each atom as shown in Figure 1.4 These lines represent **bent bonds.** Consistent with this picture, it is found that ethene is a planar molecule and that molecules of the type $XYC=CXY$, such as $HClC=CHCl$, can have two forms called **geometric isomers.** The groups X and Y are on the same side of the molecule in a *cis* isomer and on opposite sides in a *trans* isomer (Figure 1.5). Thus the subject of **stereochemistry,** the study of the shape and geometry of molecules and its relation to their properties, was born, and organic chemistry (the chemistry of carbon compounds) blossomed as chemists worked out the three-dimensional structures of thousands of carbon-containing molecules of increasing complexity just from a study of their compositions (formulas), properties, and methods of synthesis.

◆ **1.6 The Shell Model**

The first steps toward the understanding of the nature of the chemical bond could not be taken until the composition and structure of atoms had been elucidated. The model of the atom that emerged from the early work of Thomson, Rutherford, Moseley, and Bohr was of a central, very small, positively charged nucleus composed of positively charged protons and neutral neutrons, surrounded by one or more negatively charged electrons moving at high speed and effectively occupying a volume much larger than that of the nucleus. The atomic number, Z, gives the number of protons in the nucleus and the number of electrons surrounding the nucleus in a neutral atom.

The similarity in the properties of the elements in any particular group of the periodic table led to the conclusion that the atoms of the elements in a given group must have similar electron arrangements. In particular the lack of reactivity of the noble gases—no compounds of these elements were known at the time, and they were called the inert gases—led both W. Kossel (1916) and Lewis (1916) to conclude that these substances have a particularly stable arrangement of electrons. This in turn led to the development of the **shell model** of the atom. In the shell model, the electrons in an atom are arranged in successive spherical layers or shells surrounding the nucleus. The outer shell is never found to contain more than the number of electrons in the valence shell of a noble gas, namely two for helium, and eight for neon and the other noble gases. A new shell is commenced with the following element, which is an alkali metal in group 1 and has one more electron than a noble gas. Thus the arrangement of the electrons for the first 20 elements shown in Table 1.1 was deduced in which the elements in a given group have the same number of electrons in their outer shells. The shells are designated by the number n, which takes integral values starting with $n = 1$. Sometimes, following an older convention, they are designated by the letters K, L, M, N, . . . The first three shells correspond to the first three periods of the periodic table.

Table 1.1 Shell Structure of the Atoms of the First 20 Elements

Period	Z	Element	Number of Electrons in Each Shell			
			$n = 1$	2	3	4
1	1	H	1			
	2	He	2			
2	3	Li	2	1		
	4	Be	2	2		
	5	B	2	3		
	6	C	2	4		
	7	N	2	5		
	8	O	2	6		
	9	F	2	7		
	10	Ne	2	8		
3	11	Na	2	8	1	
	12	Mg	2	8	2	
	13	Al	2	8	3	
	14	Si	2	8	4	
	15	P	2	8	5	
	16	S	2	8	6	
	17	Cl	2	8	7	
	18	Ar	2	8	8	
4	19	K	2	8	8	1
	20	Ca	2	8	8	2

The outer shell is called the **valence shell** because it is these electrons that are involved in bond formation and give the atom its valence.

The completed inner shells of electrons together with the nucleus constitute the **core** of the atom. The core has a positive charge equal in magnitude to the number of electrons in the valence shell. For example, the **core charge** of the carbon atom is +4, that of the fluorine atom is +7, and that of the silicon atom is +4. The completed inner shells of electrons shield the nucleus. Thus, according to this model, the effective charge acting on the electrons in the valence shell—**the valence electrons**—is equal to the core charge. For two reasons, however, core charge is only an approximation to the actual effective charge acting on the valence shell electrons: (1) the valence shell electrons repel each other, and (2) the concept of separate successive shells is only an approximation because, as we shall see later, the shells penetrate and overlap each other to some extent. Nevertheless, for the purposes of qualitative discussion it is usually satisfactory to use the core charge.

Experimental support for the shell model has been provided by the determination of the ionization energies of free atoms in the gas phase and by the analysis of the spectra of such atoms. These measurements have given a picture of the arrangement of the electrons in an atom in terms of their energies that is essentially the same as the one we describe in Chapter 3, where we will see that this picture can also be deduced from the quantum mechanical description of an atom. Quantum mechanics also shows us that electrons do not have fixed positions in space but are in constant motion, following paths that cannot be determined. So it is strictly speaking not correct to talk about the *arrangement* of the electrons. It is only their energy, not their positions, that can be determined.

On the basis of the shell model, two apparently different models of the chemical bond were proposed, the ionic model and the covalent model.

◆ 1.7   The Ionic Model of the Chemical Bond

In 1916 Kossel noted that the loss of an electron by an alkali metal gives a positive ion, such as Na^+ (2,8) or K^+ (2,8,8), where the numbers in parentheses represent the number of electrons in successive shells. So these ions have the same electron arrangement as a noble gas. Similarly, the gain of an electron by a halogen gives a negative ion, such as a fluoride ion, F^-, (2,8) or a chloride ion, Cl^-, (2,8,8), also with the electron arrangement of a noble gas: that is, an outer shell containing eight electrons. Kossel proposed that these ions are formed because their valence shell electrons have the same stable arrangements as a noble gas. He considered solid sodium chloride to consist of positive sodium ions (cations) and negative chloride ions (anions) held together in a regular pattern by electrostatic attraction. Each crystal of solid sodium chloride can be regarded as a single giant molecule, in which a very large number of ions are arranged in a regular manner that continues through the crystal (Figure 1.6). Evidence that solids such as NaCl do consist of ions was provided by the observation that these materials are conducting in the molten state and in solution in solvents of high dielectric constant, such as water. In these states the ions are free to move independently of each other under the action of an applied electric field. Sodium chloride is a nonconductor in the solid state, because the ions are fixed in position.

Sodium chloride and many similar compounds are said to be **ionic compounds** held together by **ionic bonds.** However, even though the term "ionic bond" is widely used, it is a

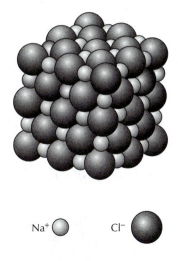

Na$^+$ ◯   Cl$^-$ ●

Figure 1.6 A space-filling model of crystalline sodium chloride.

vague and ill-defined concept. Electrostatic forces act in all directions and through relatively long distances so that the attractive forces are not confined to just two neighboring oppositely charge ions. Moreover, there are also repulsive forces between ions of like charge.

Positive alkali metal ions are easily formed because the single valence electron of an alkali metal atom is held in the atom only rather weakly by the attraction of a small core charge of +1. In other words, alkali metal atoms have a low ionization energy. The two valence electrons of a group 2 atom are also rather easily removed because they are attracted by a core charge of only +2, and so they form doubly charged ions such as Mg^{2+} and Ca^{2+} and ionic compounds such as $MgCl_2$ and CaF_2, which contain Mg^{2+} and Cl^- ions and Ca^{2+} and F^- ions respectively. The halogen atoms, each of which precedes a noble gas in the periodic table, have space in their valence shells for one more electron and, as they have a high core charge of +7, they strongly attract an additional electron to form halide ions such as F^- and Cl^-. For example, the addition of an electron to a fluorine atom is an exothermic process releasing 328 kJ mol^{-1} of energy. Similarly the elements of group 16 have room in their valence shells for two more electrons and they have a high core charge of +6 so they form doubly charged ions such as O^{2-} and S^{2-} and ionic compounds such as Na_2O and CaO. It should be noted, however, that although the addition of one electron to an oxygen atom to give the O^- ion is exothermic to the extent of 141 kJ mol^{-1}, the addition of a second electron is an endothermic process absorbing 744 kJ mol^{-1}, so that the overall process $O + 2e \rightarrow O^{2-}$ is also endothermic to the extent of 603 kJ mol^{-1}. An isolated oxide ion is therefore unstable and spontaneously loses an electron, but it is stabilized in an ionic crystal by the additional energy released when oppositely charged ions pack together to give a crystal. Indeed this energy, called the **lattice energy,** makes an important contribution to the stability of all ionic crystals.

The structures of ionic crystals are determined mainly by the ways in which oppositely charged ions of different sizes and different charges can pack together to minimize the total electrostatic energy. The sizes of ions are discussed in Chapter 2. Structures of some typical ionic crystals are given in Figure 1.7. In this figure the structures, expanded so that the ions are no longer touching, are connected by lines that serve to emphasize the geometric arrangement of the ions.

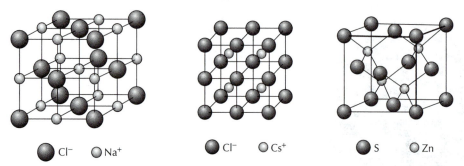

Figure 1.7 The structures of crystalline sodium chloride (NaCl), cesium chloride (CsCl), and zinc sulfide (ZnS).

Although the ionic model has been used almost exclusively to describe the bonding in a large class of solids with infinite three-dimensional structures consisting of oppositely charged ions, in which each crystal can be regarded as a giant molecule, the bonding in other much smaller molecules may also be ionic, as we shall discuss later. A simple example is provided by molecules such as NaCl and $MgCl_2$, which are formed from solid sodium and magnesium chlorides when they vaporize at high temperatures. To indicate their ionic nature, they may be written as Na^+Cl^- and $Cl^-Mg^{2+}Cl^-$.

◆ 1.8 Covalent Bonds and Lewis Structures

Clearly the explanation of the chemical bond given by Kossel cannot apply to homonuclear molecules such as Cl_2. Almost simultaneously with the publication of Kossel's theory, Lewis published a theory that could account for such molecules. Like Kossel, Lewis was impressed with the lack of reactivity of the noble gases. But he was also impressed by the observation that the vast majority of molecules have an even number of electrons, which led him to suggest that in molecules, electrons are usually present in pairs. In particular, he proposed that in a molecule such as Cl_2 the two atoms are held together by sharing a pair of electrons because in this way each atom can obtain a noble gas electron arrangement, as in the following examples:

$$:\ddot{C}l:\ddot{C}l: \qquad H:H$$

Diagrams: of this type are called **Lewis diagrams** or **Lewis structures.** The bond between the two atoms could be called a shared-electron-pair bond but it is now universally called a **covalent bond**—a term introduced by Irving Langmuir (1919). In drawing Lewis structures, the core of the atom is represented by the symbol of the element and the valence shell electrons by one to eight dots, the first four arranged singly around the symbol for the core, with additional electrons used to form pairs as follows:

Valence	1	2	3	4	3	2	1	0
	·H							He:
	·Li	·Be·	·B·	·Ċ·	·N̈·	:Ö·	:F̈·	:N̈e:
	·Na	·Mg·	·Al·	·S̈i·	·P̈·	:S̈·	:C̈l·	:Är:

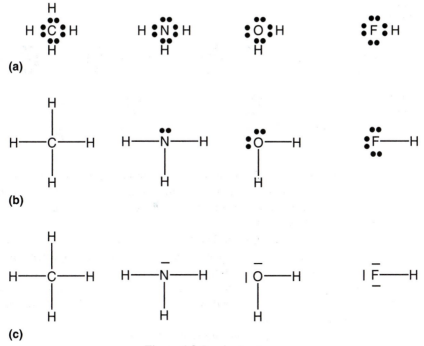

Figure 1.8 Lewis structures.

The complete symbol for each element can be called its **Lewis symbol.** The number of unpaired electrons in the symbol equals the number of bonds that the atom can form, that is, its valence. Each unpaired electron can be paired with an unpaired electron in the Lewis symbol of another element to form a shared pair or covalent bond. In this way the atoms of the elements in groups 14–17, such as C, N, O and F, can attain a noble gas electron arrangement as shown by the Lewis structures in Figure 1.8a. The elements in groups 1, 2, and 13 such as Li, Be, and B do not, however, achieve a noble gas electron arrangement even when they form the maximum number of bonds (see Section 1.13). A covalent bond (a shared electron pair) is usually designated by a bond line rather than by a pair of dots (Figure 1.8b). As we noted earlier, and as we will discuss in detail later, some elements have more than one valence. The valence given by the number of unpaired electrons in the Lewis symbol for an element, as illustrated above, is called its **principal valence.**

In a Lewis diagram, the pairs of electrons that are not forming bonds are called **nonbonding pairs** or more usually **lone pairs.** A lone pair is usually designated by a pair of dots but less commonly by a single line (Figure 1.8c). In the Lewis diagrams for the CF_4, NF_3, OF_2, and F_2 molecules (Figure 1.9) each fluorine atom has three lone pairs, oxygen two, and nitrogen one.

Lewis called the apparent tendency of atoms to acquire a noble gas electron arrangement, either by forming ions or by sharing electron pairs, the **rule of eight.** Later Langmuir called it the **octet rule,** and this is the term that is now generally used. Lewis did not regard the rule of eight as being as important as the **rule of two,** according to which electrons are

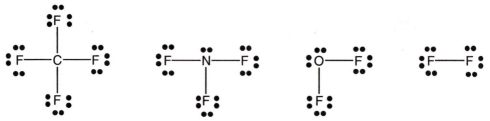

Figure 1.9 Lewis structures of some fluorides.

present in molecules in pairs (Box 1.2), because he found more exceptions to the octet rule than to the rule of two. There are only a few exceptions to the rule of two, such as molecules with an odd number of electrons (free radicals), whereas there are a large number of exceptions to the octet rule (Section 1.13).

Because CX_4 molecules have a tetrahedral geometry, Lewis postulated that the four pairs of electrons in the valence shell of the carbon atom have a tetrahedral arrangement, thus giving the four covalent bonds a tetrahedral geometry. Later, when the angular geometry of the OX_2 molecules and the pyramidal geometry of NX_3 molecules were established, it became clear that the directed nature of covalent bonds in many molecules could be rationalized on the basis of the tetrahedral arrangement of four pairs of electrons in the valence shell of an atom (Figure 1.10). In contrast, ionic bonds are said to be nondirectional because Coulomb

▲ BOX 1.2 ▼
Lewis and the Electron Pair

Although Lewis had no clear idea of why electrons are found in molecules as pairs, or how a shared pair of electrons holds two atoms together, the ideas of the shared electron pair—the covalent bond—and the octet rule enable us to understand the formulas of a vast number of molecules and their relationship to the positions of the elements in the periodic table. Because the formation of electron pairs seemed to contradict Coulomb's law, according to which electrons repel each other so that they should keep as far apart as possible, Lewis even suggested that Coulomb's law is not obeyed over the very short distances between electrons in atoms and molecules. Although we now know that Coulomb's law is obeyed for all distances between charges, in making the assumption about the importance of electron pairs, Lewis displayed remarkable intuition: electrons do indeed form pairs in most molecules, despite their mutual electrostatic repulsion. We now have much a much more detailed and exact knowledge about the distribution of the electrons in molecules than is given by Lewis diagrams, but Lewis diagrams showing bonding pairs and lone pairs are still widely used today, and the electron pair remains a central concept in chemistry.

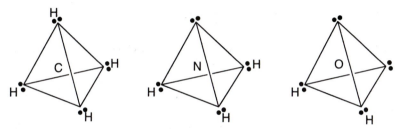

Figure 1.10 The tetrahedral, trigonal pyramidal, and angular geometries of the methane, ammonia, and water molecules based on the tetrahedral arrangement of four electron pairs.

forces act in all directions. So the arrangement of anions around a cation in an ionic crystal or molecule is not determined by the arrangement of electron pairs in the valence shell of the cation but by the geometry that enables anions to pack as closely as possible around the cation, thus decreasing the potential energy of the crystal.

As we have seen, some atoms, such as carbon, oxygen, and nitrogen, form double and triple bonds. Lewis represented these bonds as consisting of two and three shared pairs, respectively (Figure 1.11). Since the four pairs in an octet have a tetrahedral arrangement, a double bond can be represented by two tetrahedra sharing an edge and a triple bond by two tetrahedra sharing a face. These models agree with the observed planar geometry of ethene and related molecules and the linear geometry of ethyne and related molecules (Figure 1.12).This model is similar to the bent-bond models in Figure 1.4 in that the tetrahedral arrangement of bonds or electron pairs around each atom is maintained.

Figure 1.11 Lewis structures of ethene and ethyne.

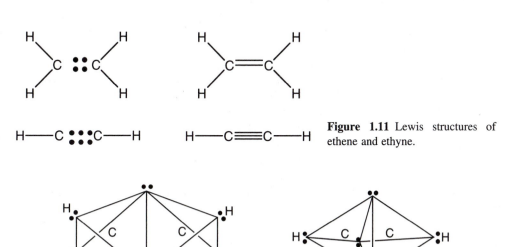

Figure 1.12 Structures of ethene and ethyne, based on the tetrahedral arrangement of four electron pairs around each carbon atom.

◆ 1.9 Polar Bonds and Electronegativity

Ionic bonds and covalent bonds appear, at first sight, to be of two completely different kinds. However, Lewis maintained that there was no fundamental difference between them. He recognized that a shared electron pair is generally not shared equally between the two bonded atoms unless they are atoms of the same kind. The atoms of the elements on the right side of the periodic table attract electrons into their valence shells more strongly than those on the left because they have higher core charges. Thus in a molecule such as H—Cl, the chlorine atom acquires a greater "share" of the bonding electron pair than the hydrogen atom. In effect it acquires more than an equal share of two electrons (more than the one electron that would give it a zero charge but fewer than two), so it has a resulting small negative charge, leaving the hydrogen atom with an equal and opposite small positive charge. The bond between the two atoms is then called a **polar covalent bond,** or simply a **polar bond.** We might depict a nonpolar "pure covalent" bond by placing the shared pair midway between the two bonded atoms and a polar covalent bond by placing the shared pair closer to the atom that has the larger share of the pair. However, this not is a particularly convenient or

---

▲ BOX 1.3 ▼
Bond Lines

There has never been a really clear understanding of what a bond line stands for. Originally it was meant to indicate simply that the two atoms between which it is drawn are held strongly together. However, it is now usually taken to indicate a shared pair of electrons, that is, a covalent bond. In contrast, the presence of ionic bonds in a molecule or crystal is usually implied by the indication of the charges on the atoms, and no bond line is drawn. This immediately raises the question of how polar a bond has to be before the bond line is omitted. Whereas the structure of the LiF molecule would normally be written as Li^+F^- without a bond line, even the highly ionic BeF_2 is often written as F—Be—F rather than as F^- Be^{2+} F^-.

Even though it is well known that the bonds in these molecules are polar, writing their structures with bond lines gives the impression that the bonding is predominately covalent. However, omitting these lines for predominately ionic molecules leads to difficulty because it is then harder to clearly indicate their geometry. The solution to this problem is not obvious, but we need to be aware that a bond line does not necessarily imply a predominately covalent bond. In many ways it would be simplest to return to the original use of a bond line, namely, to indicate that two atoms that are bonded together, whether the bonding is predominately covalent or predominately ionic.

Finally, we should note that the lines that are often drawn in illustrations of three-dimensional ionic crystal structures to better show the relative arrangement of the ions do not represent shared pairs of electrons, that is, they are not bond lines.

generally useful representation, and a polar bond is usually represented by a bond line sometimes with the symbols $\delta+$, representing a small positive charge ($0 < \delta < 1$), and $\delta-$, representing a small negative charge, added to the appropriate atoms (Box 1.3).

$$H^{\delta+}—Cl^{\delta-}$$

In 1932 Pauling introduced the term **electronegativity** to describe

the power of an atom in a molecule to attract electrons to itself.

In general, metallic elements have low electronegativities—that is, they attract electrons only
weakly—while nonmetals have high electronegativities—that is, they attract electrons
strongly because they have high core charges. Because electronegativity is not defined in a
quantitative way it is, strictly speaking, not possible to assign a quantitative value for the
electronegativity of the atoms of an element. Nevertheless several attempts have been made
to devise quantitative scales that express the relative electronegativities of the elements. The
original scale is due to Pauling, who based it on the difference in the dissociation energy of
an AB molecule and the average of the dissociation energy of the A_2 and B_2 molecules. Mulliken based his scale on the average of the ionization energies and electron affinities of an
atom, while Allred and Rochow (1958) proposed a scale based on the force exerted on a
electron in the valence shell of an atom, which they took to be $Z_{eff}e^2/r^2$ where Z_{eff} is the effective nuclear charge, e is the unit of electric charge, and r is the covalent radius. We define "covalent radius" in Chapter 2, but essentially it is the size (radius) of an atom in the
bond direction. Still other scales have been proposed, but it is not possible to choose any one
of these scales as being superior to the others because they are all defined in different ways,
none of which is the same as the qualitative definition given by Pauling. However, rather
surprisingly perhaps, considering the very different basis of each of the scales, they give
comparable relative values, so that when adjusted to cover the same range as the Pauling
values, they give similar values. So almost any of these scales is useful for making an approximate comparison of the electronegativities of the elements. Table 1.2 gives the set of
values due to Allred and Rochow. We quote these values to two significant figures only because there is no justification for using more precise values. Despite its qualitative nature,
the concept of electronegativity has proved very useful in the development of our ideas concerning the chemical bond. The most important use of electronegativity values is to estimate
the polarity of bonds, that is, to obtain rough estimates of the charges on atoms in molecules.
Various theoretical methods have been proposed for calculating atomic charges, but they
give substantially different results because until recently, there has been no sound definition
of atomic charge and therefore, of course, no way of determining it experimentally. In Chapter 6 we discuss how atomic charge can be clearly defined in terms of the electron density,
which can be both calculated and also determined experimentally by X-ray crystallography.

It is important to point out that almost all bonds are polar bonds, whether they are approximately described as covalent or ionic. The bonds in the molecules of the various forms
of the elements such as the diatomic molecules H_2, Cl_2, and N_2, larger molecules such as P_4
and S_8, and infinite molecules such as diamond may be described as "pure covalent" bonds

Table 1.2 Electronegativity Values According to Allred and Rochow

Period	Group							
	1	*2*	*13*	*14*	*15*	*16*	*17*	*18*
1	H							He
	2.2							—
2	Li	Be	B	C	N	O	F	Ne
	1.0	1.5	2.0	2.5	3.1	3.5	4.1	—
3	Na	Mg	Al	Si	P	S	Cl	Ar
	1.0	1.2	1.3	1.7	2.1	2.4	2.8	—
4	K	Ca	Ga	Ge	As	Se	Br	Kr
	0.9	1.0	1.8	2.0	2.2	2.5	2.7	3.1
5	Rb	Sr	In	Sn	Sb	Te	I	Xe
	0.9	1.0	1.5	1.7	1.8	2.0	2.2	2.4

because the bonding electrons are necessarily shared equally and the atoms have a zero charge. The C—C bonds in ethane (H_3C—CH_3) and the N—N bond in hydrazine (H_2N—NH_2) are also pure covalent bonds because the carbon and nitrogen atoms are completely equivalent and therefore attract the bonding electrons equally strongly. Even in a molecule such as chloroethane (ClH_2C—CH_3), however, the two carbon atoms are not exactly equivalent and do not therefore attract electrons equally strongly—they have slightly different electronegativities—and so the two carbon atoms have different small charges and the CC bond has a small polarity. Such a bond is said to have a large covalent character and a small ionic character. Conversely, when the difference in electronegativity of the bonded atoms is large, the atoms are expected to have large charges and the bond between them may be regarded as having a large ionic character. There are no "pure ionic" bonds because there is always at least a small amount of sharing of electrons between any two ions. Although the terms "ionic character" and "covalent character," like "electronegativity," are widely used, they cannot be quantitatively defined and so their meaning is not entirely clear. The uncertainty in the exact meaning of these terms has led to misunderstanding and controversy in discussions of bonding. We return to the determination of the charges of atoms in molecules and the concepts of ionic and covalent character in Chapters 6, 8, and 9.

We note in passing that two atoms of the same element in a molecule, such as the two carbon atoms in CH_3CH_2Cl, may have slightly different electronegativities. As a result, it is, strictly speaking, not possible to assign a fixed constant value for the electronegativity of an atom, which is another reason for giving the values in Table 1.2 to only two significant figures.

That the geometry of a covalent molecule is determined by the directional character of the bonds whereas the geometry of an ionic crystal or molecule is determined by the packing of negative ions around a positive ion raises questions such as: What determines the geometry of a polar covalent molecule? How directional is a polar covalent bond? Is the planar geometry of the BCl_3 molecule, in which the bonds are very polar, due to the directional character of the B—Cl bonds or to the packing of an anion-like negatively charged Cl atoms around a cation-like boron atom? We return to these questions in later chapters.

◆ 1.10   Polyatomic Ions and Formal Charge

Polyatomic ions are groups of atoms that are held strongly together as in a molecule but have an overall positive or negative charge. In other words, they are charged molecules. They are found in ionic crystals in association with an ion of opposite charge. For example, ammonium chloride, NH_4Cl, consists of polyatomic NH_4^+ ions (ammonium: Figure 1.13a) and chloride ions, and sodium tetrafluoroborate, $NaBF_4$, consists of polyatomic BF_4^- ions (tetrafluoroborate: Figure 1.13b) and sodium ions (Figure 1.13). The recognition of polyatomic ions solved the problem of representing many of the so-called molecular compounds that we mentioned in Section 1.4, such as $2KF \cdot SiF_4$, which contains the polyatomic ion SiF_6^{2-} and is therefore more correctly formulated as $(K^+)_2 \, SiF_6^{2-}$.

In the Lewis diagram for a polyatomic ion the charge is often allocated specifically to one of the atoms on the assumption that each bonding pair of electrons is shared equally between the two bonded atoms: that is, on the assumption that the bonding is purely covalent. In the ammonium ion, four electrons, one from each bond, are allocated to the nitrogen atom which, since it needs five electrons to balance its core charge of $+5$, has a resultant single positive charge. One electron is allocated to each hydrogen atom, which is just sufficient to balance the nuclear charge of $+1$, giving a resultant zero charge (Figure 1.14). In the tetrafluoroborate ion, four electrons, one from each bond, are allocated to the boron atom, which, since it needs only three electrons to balance its core charge of $+3$, has a resultant charge of -1. One electron is allocated to each fluorine atom, giving a resultant zero charge. It is also necessary to allocate charges to atoms in some neutral molecules in order to write structures that obey the octet rule, for example, as in trimethylamine oxide $(CH_3)_3NO$ and the molecule F_3BNH_3 (Figure 1.14).

The charges allocated in this way are called **formal charges.** They do not in general show the actual charge distribution in a molecule or ion because of the polarity of most bonds. *Formal charges may even be of opposite sign to the real charge.* For example, the boron atom in BF_4^- has a formal negative charge but, as we shall see later, because of the high electronegativity of fluorine, the real charge on boron is positive. *The concept of formal charge is useful only for the purpose of the keeping track of electrons when one is writing Lewis structures that do not take account of bond polarity.*

A nitrogen atom can form four bonds only if it loses an electron to become N^+ so that it is then isoelectronic with a carbon atom. **Isoelectronic** atoms or molecules have the same number of valence electrons, arranged in the same way. Thus B^-, C, and N^+ are isoelectronic atoms and can each form four bonds. Some examples of isoelectronic molecules are illustrated in Figure 1.15.

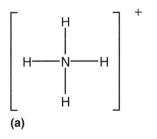

(a)

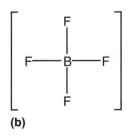

(b)

Figure 1.13 Lewis structures of (a) the ammonium ion NH_4^+ and (b) the tetrafluoroborate ion BF_4^-.

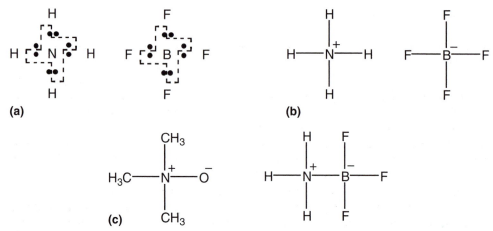

(a) **(b)**

(c)

Figure 1.14 Formal charges: assigning one electron of each bonding pair to each of the bonded atoms (a) leads to the formal charges in (b). Formal charges in some neutral molecules are shown in (c).

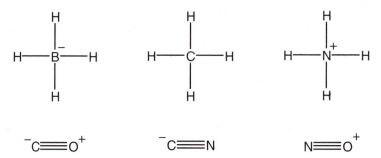

Figure 1.15 Two sets of isoelectronic molecules.

◆ 1.11 Oxidation Number (Oxidation State)

Polyatomic ions illustrate one of the difficulties with the concept of valence as we have defined it. Boron, normally considered to have a valence of 3 because, for example, it forms three bonds in molecules such as BCl_3, and four bonds in BCl_4^-. Is its valence then 4? Should we assign a valence of 3 to boron only when it has a formal zero charge and a valence of 4 to boron when it has a negative formal charge? Difficulties such as this have led to the replacement of the concept of valence, particularly for the description of inorganic compounds, by the concept of **oxidation number,** or **oxidation state.** The oxidation number of an atom in a molecule is defined as the charge the atom would have if both the electrons in any bond that it forms are transferred to the more electronegative of the two atoms, in other words, as if the molecule were formulated as ionic. Thus boron in both BCl_3 and BCl_4^- has an oxidation number of $+III$ and chlorine an oxidation number of $-I$, while nitrogen in both NH_3 and NH_4^+ has an oxidation number of $-III$ and hydrogen an oxidation number of $+I$. Roman numerals are usually used for oxidation numbers to distinguish them from charges. Oxidation numbers are also convenient for the description of the molecules of elements that have several valences, such as sulfur. For example, the sulfur atom in SO_2 is in the $+IV$ oxidation state whereas in SO_3 it is in the $+VI$ oxidation state. In contrast to inorganic compounds, which frequently have considerable ionic character, oxidation numbers are not very useful for carbon compounds, which are predominately covalent and for which the constant tetravalence of carbon is one of the cornerstones of organic chemistry.

Formal charge and oxidation number are two ways of defining atomic charge that are based on the two limiting models of the chemical bond, the covalent model and the ionic model, respectively. We expect the true charges on atoms forming polar bonds to be between these two extremes.

◆ 1.12   Donor–Acceptor Bonds

Ammonia reacts with boron trichloride to form a molecule called an **adduct** or **Lewis acid base complex** in which the lone pair on the ammonia molecule is shared with the boron atom to form a covalent bond and completing an octet on boron (Figure 1.16):

$$H_3N: + BCl_3 \rightarrow H_3N\text{---}BCl_3$$

We should note that the formation of this bond confers formal charges on the B and N atoms. In this bond and many similar Lewis acid–base complexes both the electrons forming the bond come from the same atom rather than from different atoms, as in the formation of a bond between two chlorine atoms. This type of bond is often called a **donor–acceptor bond,** a **dative bond,** or a **coordinate bond,** and is sometimes given a special symbol—an arrow denoting the direction in which the electron pair is donated:

$$H_3N \rightarrow BCl_3$$

Molecules of this type are often called **donor–acceptor complexes** or sometimes **charge transfer complexes** (because charge is transferred from the donor to the acceptor as the nonbonding electron pair of the donor atom is shared with the acceptor atom). In other words, there is a formal transfer of one electron, which is evident in the formal charges on the atoms in the complex. Once formed, however, the bond is simply a covalent bond consisting of a pair of shared electrons, whose origin is irrelevant to the nature of the

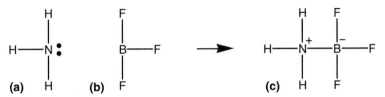

Figure 1.16 The ammonia–boron trifluoride donor–acceptor complex: (a) donor Lewis base, (b) acceptor Lewis acid, (c) the donor–acceptor or Lewis acid–base complex.

bond because all electrons are identical. Thus, although the concept of donor and acceptor molecules is useful, a special name and symbol for the bond formed between them is
not really necessary. Although there is no difference between a coordinate covalent bond
and a "normal" covalent bond in molecules in their equilibrium geometry, a difference
becomes evident when the bond is broken. Breaking a bond in a Cl_2 molecule gives two
Cl atoms

$$:\ddot{\text{C}}l:\ddot{\text{C}}l: \rightarrow :\ddot{\text{C}}l\cdot + :\ddot{\text{C}}l\cdot$$

In contrast breaking the bond in the $H_3N:BCl_3$ molecule gives two stable molecules $H_3N:$
and BCl_3. In the first case the bond breaks symmetrically while in the second case it breaks
unsymmetrically.

◆ 1.13 Exceptions to the Octet Rule: Hypervalent and Hypovalent Molecules

Lewis recognized that certain molecules such a PCl_5 and SF_6 are exceptions to the octet rule
because their Lewis structures indicate that the central atom has more than eight electrons
in its valence shell: 10 for the P atom in PCl_5 and the S atom in SF_4, and 12 for the S atom
in SF_6 (Figure 1.17). Such molecules are called **hypervalent** because the valence of the central atom is greater than its principal valence. To write a Lewis structure for such molecules,
the Lewis symbol for the hypervalent atom must be modified to show the correct number of
unpaired electrons. For the molecules in Figure 1.17 we would need to write the Lewis symbols as follows:

$$\cdot\dot{\text{P}}\cdot \qquad \cdot\ddot{\text{S}}\cdot \qquad \cdot\ddot{\text{S}}\cdot$$

Hypervalent molecules are relatively common for the elements of period 3 and beyond. It is
often said that they are formed only by the most electronegative ligands, in particular, F, Cl,
$=$O, and OX, with the nonmetals of period 3 and subsequent periods. But in many cases the
ligand atom attached to the central atom is carbon, as in $As(CH_3)_5$, and $P(C_6H_5)_5$, in which
the electronegativity of the central carbon atom (2.5) is only slightly greater than that of either arsenic (2.2) or phosphorus (2.1). We will see later that the relative sizes of the central
atom and the ligand atoms are important in determining the occurrence of hypervalent molecules, because these differences in size allow more than four such ligands to be packed
around a sufficiently large central atom.

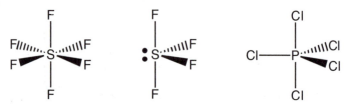

Figure 1.17 Some examples of hypervalent molecules that have more than eight electrons in the valence shell of the central atom.

Because the octet rule had proved so useful for understanding and describing the bonding in so many molecules, and because this rule came to be regarded more as a law than as a summary of observations, the bonding in hypervalent molecules has often been considered to be in some way different from that in "ordinary" molecules that obey the octet rule. Despite the later discovery of the noble gas compounds (Box 1.4) and the preparation of many other hypervalent molecules whose properties do not differ significantly from analogous non-hypervalent (octet rule) molecules, it is still often believed that there is something abnormal about the bonding in these molecules. The bonding in hypervalent molecules has been formulated in terms of several different models to avoid violating the octet rule. There has been considerable controversy concerning the relative merits of these models, which we will discuss in later chapters. We will see that much of this controversy has arisen as a consequence of a lack of appreciation of the limitations of Lewis structures and an overemphasis on the octet rule, and indeed no special descriptions of the bonding in hypervalent molecules are necessary.

▲ BOX 1.4 ▼
The Octet Rule and the Noble Gases

Although the octet rule was first formulated on the basis of the observed lack of reactivity of the noble gases, and the observation that in many molecules each atom has eight electrons in its valence shell, it was often cited in later years as a reason for the absence of any known compounds of the noble gases. This acceptance of the octet rule as a law of nature rather than as an empirical rule even inhibited the continued search for compounds of the noble gases after the initial failure of Moissan, in 1895, to find any conditions under which fluorine, which he had discovered in 1886, would react with a sample of argon provided by Ramsay, who first identified argon. Consequently it came as a great surprise to most chemists when the first noble gas compound, $XePtF_6$ was prepared in 1962 by Bartlett. Pauling, however, was one of the few chemists who were not surprised. In the 1930s he had predicted, mainly on the basis of the existence of molecules such as BrF_5, IF_7, and H_5IO_6, that it should be possible to prepare analogous compounds of xenon including fluorides such as XeF_6. He persuaded his colleagues Yost and Kaye to attempt the preparation of this compound, by the reaction of xenon and fluorine. Unfortunately they were unsuccessful. Although they may well have prepared a very small amount of a xenon fluoride, they were unable to show this definitively. Subsequently there appears to have been little interest in trying to repeat this experiment. So it continued to be generally accepted that compounds of the noble gases could not be prepared until Bartlett prepared $XePtF_6$ by the reaction between PtF_6 and xenon. This discovery was followed rapidly by the preparation of a variety of fluorides, oxides, and oxofluorides of xenon, such as XeF_4, XeO_3, and $XeOF_4$. Since then compounds of krypton, such as KrF_2, as well as compounds with Xe—N and Xe—C bonds, have also been prepared. All these molecules are necessarily hypervalent.

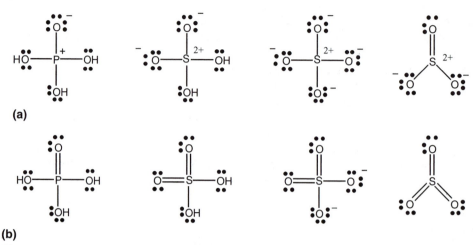

(a)

(b)

Figure 1.18 Lewis structures of oxo acids and oxides of phosphorus and sulfur: (a) octet rule structures according to Lewis and (b) hypervalent structures.

Many common and well-known molecules such as the oxides and oxoacids of sulfur and phosphorus in their higher oxidation states (e.g., SO_2, SO_3, H_2SO_4, H_3PO_4) must be regarded as hypervalent if they are described by their classical structural formulas in which the bonds to oxygen are double bonds (Figure 1.18). However, Lewis drew his diagrams for these molecules so that they obeyed the octet rule with a formal negative charge on oxygen and a corresponding formal charge on P, or S, although this was inconsistent with his recognition of molecules such as PF_5 and SF_6 as exceptions to the octet rule, and these octet rule structures have been widely adopted.

There are also molecules that are exceptions to the octet rule because one of the atoms has fewer, rather than more than, eight electrons in its valence shell in the Lewis structure (Figure 1.19). These molecules are formed by the elements on the left-hand side of the periodic table that have only one, two, or three electrons in their valence shells and cannot therefore attain an octet by using each of their electrons to form a covalent bond. The molecules LiF, $BeCl_2$, BF_3, and $AlCl_3$ would be examples. However, as we have seen and as we will discuss in detail in Chapters 8 and 9, these molecules are predominately ionic. In terms of a fully ionic model, each atom has a completed shell, and the anions obey the octet rule. Only if they are regarded as covalent can they be considered to be exceptions to the octet rule. Covalent descriptions of the bonding in BF_3 and related molecules have therefore

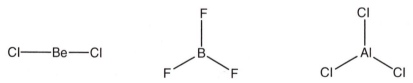

Figure 1.19 Some examples of molecules that are exceptions to the octet rule because the central atom has fewer than eight electrons in its valence shell.

been devised so that they appear to obey the octet rule, but we shall see later that these special descriptions are unnecessary.

Molecules such as $BeCl_2$, BF_3, and $AlCl_3$, which have space in their valence shells for one or two more electron pairs and in which the central atom is positively charged, are good acceptor molecules or Lewis acids (Section 1.12), forming polyatomic ions such as BF_4^- and $AlCl_4^-$ and donor–acceptor complexes such as $BeCl_2(OEt_2)_2$ and $BF_3 \cdot NH_3$.

We should note that hydrogen never has more than two electrons in its valence shell in the Lewis diagram of any of its molecules because its valence shell is filled by just two electrons. Thus the octet rule is not applicable to hydrogen.

◆ 1.14 Limitations of the Lewis Model

Lewis structures, according to which the valence shell electrons in a molecule are arranged in bonding and nonbonding pairs, have played a very important role in the development of our understanding of the chemical bond, and indeed they still form a most useful basis for the discussion of the properties of molecules. However, they have many limitations. We have already noted that they do not provide a very convenient representation of molecules in which the bonds are polar and that they are not useful for molecules in which the bonding is predominately ionic. Moreover, many molecules are exceptions to the octet rule, which has been incorporated into the Lewis model even though Lewis himself recognized its limitations. And there are molecules, such as the boranes, in which the bonding cannot be described in terms of localized electron pairs. In the following chapters we will encounter other limitations, and we will see that many controversies about bonding have arisen because of a failure to understand and recognize the limitations of Lewis structures.

However, there are more serious problems. A Lewis structure provides a static model of the electron distribution, yet a fundamental theorem of electrostatics states that no system of charges can be at equilibrium while the charges are at rest. A more realistic description of the electron distribution must take into account the motion of the electrons and their wavelike nature. In Chapter 3 we will see that the distribution of the electrons in atoms and molecules cannot be described in classical terms but only in terms of quantum mechanics, according to which we can determine no more than the probability of finding an electron at a given point. Thus we describe the distribution of the electrons by a distribution of probability density, which can be conveniently represented as a cloud of negative charge. We will see why, nevertheless, the electron pair plays such a dominant role in the electronic structure of molecules and why the picture of precisely located electron pairs provided by a Lewis structure is so useful, even though only the average distribution of the electrons can be determined.

▶ References

A. L. Allred and E. G. Rochow, *J. Inorg. Nucl. Chem. 5,* 264, 1958.
G. N. Lewis, *J. Am. Chem. Soc. 38,* 762, 1916.
I. Langmuir, *J. Am. Chem. Soc. 41,* 868, 1919.
W. Kossell, *Ann. Phys. (Leipzig) [4], 49,* 229, 1916.

▶ Further Reading

N. Bartlett and D. H. Lohmann, *J. Chem. Soc.* 5253, 1962.
The preparation of $XePtF_6$—the first noble gas compound.
P. L. Laszlo and G. J. Schrobilgen, *Angew. Chem. Int. Ed. Engl. 27,* 479, 1988.
An interesting history of the discovery of noble gas compounds.
L. Pauling, *J. Am. Chem. Soc. 55,* 1895, 1933.
Prediction of XeF_6 and other noble gas compounds.
D. M. Yost and A. L. Kaye, *J. Am. Chem. Soc. 35,* 3052, 1933.
Attempted preparation of a xenon fluoride.

C H A P T E R

2

BOND

PROPERTIES

■ ■ ■

◆ 2.1 Introduction

In Chapter 1 we discussed the origin and early development of the concept of the chemical bond. With the subsequent development of X-ray crystallography, electron diffraction, and various spectroscopic techniques, it became possible for the first time to obtain quantitative structural information on molecules and crystals, hence on their bonds. An enormous amount of such information has been accumulated by these methods over the past 80 years. We can measure the distances between the atomic nuclei in a molecule and thus obtain the bond lengths, as well as the angles between bonds (bond angles and torsional angles). These are the only well-defined properties of bonds that can be accurately determined unambiguously for any polyatomic molecule. Consequently bond lengths and bond angles have played a prominent role in the discussion of the nature of the chemical bond. And this information is now being supplemented by data obtained from high-level ab initio calculations (Chapter 6), which in many cases can now give values comparable to those obtained by experimental methods. Moreover, these calculations can give us information on molecules that have not yet been prepared or had their structure determined experimentally. This information is often particularly valuable for comparison with known molecules. The major part of this chapter is devoted to bond lengths and their interpretation to give information about the nature of bonds.

An important related property of a bond is its strength. The strength of a bond in a molecule can be measured by the stretching force constant, obtained either from the vibrational spectrum of a molecule or by the dissociation energy obtained from the electronic spectrum or, most often, from thermochemical measurements. However, accurate stretching force constants can be obtained for diatomic molecules only because none of the bonds in a polyatomic molecule vibrate independently of the others. The vibrational spectrum of a polyatomic molecule can be analyzed by a method called normal coordinate analysis, but this does not necessarily give such reliable or accurate force constant values as can be obtained from a diatomic molecule. Similarly accurate bond dissociation energies can be obtained only for diatomic molecules because breaking one bond in a polyatomic molecule affects the strength of all the neighboring bonds. As we shall see, there is usually a good correlation between bond length and bond strength: *in general, the shorter the bond between two given atoms, the stronger it is.*

The relationships between bond length, stretching force constant, and bond dissociation energy are made clear by the **potential energy curve** for a diatomic molecule, the plot of the change in the internal energy ΔU of the molecule A_2 as the internuclear separation is increased until the molecule dissociates into two A atoms:

$$A_2 \rightarrow 2A$$

A typical potential energy curve for a diatomic molecule in its ground state is shown in Figure 2.1. Considering the reverse process, namely, the formation of the A_2 molecule from two A atoms, we see that the energy of the molecule decreases as the two atoms approach and the bond begins to form, as the attraction between the bonding electrons and the nuclei increases. As the nuclei approach each other, the repulsion between them increases and eventually becomes sufficiently great that the total energy of the molecule passes through a minimum and begins to increase.

The minimum of the potential energy curve occurs at the **equilibrium bond length, r_e,** of the molecule. The depth of the minimum is the change in the electronic contribution to the internal energy ΔU_{el} for a hypothetical state of the molecule at 0 K that has no vibrational, rotational or translational energy (i.e., the energy obtained from ab initio calculations). The deeper the minimum, the more strongly the atoms are bonded together. For the hydrogen molecule, $\Delta U_{el} = 458$ kJ mol^{-1}:

$$H_2 \rightarrow 2H \qquad \Delta U_{el} = 458 \text{ kJ mol}^{-1}$$

At 298 K ΔU includes vibrational, rotational, and translational energy changes that total 25 kJ mol^{-1}, of which the most important is the vibrational energy, so that the quantity ΔU_{298} that is measured at 298 K is

$$\Delta U_{298} = \Delta U_{el} - \Delta U_{vib, rot, trans} = 458 - 25 = 433 \text{ kJ mol}^{-1}$$

This is the quantity called the **bond dissociation energy** or **bond energy.**

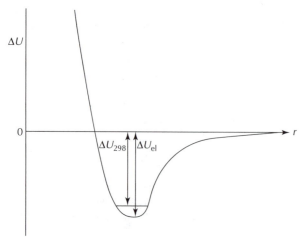

Figure 2.1 Plot of the energy change ΔU for the dissociation of a diatomic molecule. ΔU_{el} is the value for the hypothetical state of the molecule at 0K that has no vibration, rotational or translational energy. ΔU_{298} is for the value for the dissociation of the molecule at 298K and includes vibrational, rotational and translational energy changes.

The slope (gradient) of the curve on either side of the minimum shows how rapidly the energy of the molecule rises as the bond is stretched or compressed, hence it governs the force constant of the bond and (in combination with the masses) the vibrational frequency of the bond. The steeper the curve on either side of the minimum, the greater the force constant and (for given masses) the higher the vibrational frequency. A deep minimum usually has steep sides so that a molecule with a large dissociation energy usually has a large force constant, and vice versa. However, it should be realized that the force constant is a curvature rather than a slope; that is, it is a second derivative of the energy with respect to displacement. For example, the potential of the harmonic oscillator is a parabola with the equation $V = 1/2kx^2$, and the larger the force constant k, the more curved the parabola becomes.

Another important property of a molecule is its electric dipole moment. A molecule has an electric dipole moment when the center of positive charge resulting from the nuclear charges does not coincide with the center of negative charge due to the electrons. It is therefore a function of the bond lengths and angles and the electron distribution. It is, strictly speaking, not a bond property, although we may think of each bond as having a bond dipole that contributes to the overall dipole moment.

We discuss bond lengths in the next section, but we defer the discussion of bond angles to Chapters 4 and 5, where we discuss all aspects of molecular geometry. In later sections of this chapter we discuss bond strength in terms of bond enthalpies and force constants, the determination of approximate values for these properties in polyatomic molecules, and the determination and analysis of dipole moments.

◆ 2.2 Bond Lengths and Covalent Radii

The single most well-defined property of a chemical bond in a molecule is its length—the distance between the nuclei of the two atoms that are bonded together—called the **bond length.** However, it is important to realize that the experimentally measured length of a bond is only an average value that has some uncertainty because of molecular vibrations and rotations. Moreover, different experimental techniques do not measure quite the same parameter. Electron diffraction gives the distance between two nuclei, but X-ray crystallography gives the distance between the peaks of maximum electron density that are very close to but not necessarily exactly at the position of the nucleus. Finally we should note that an experimentally measured bond length is also necessarily slightly different from an ab initio calculated bond length, which is the distance between two hypothetically motionless nuclei in a free molecule. This distance is called the equilibrium bond length. We use "hypothetical" because there is no motionless molecule in reality. Even at 0 K, all molecules possess a certain amount of energy, the zero-point energy of the ground vibrational state, and therefore all the atoms have some motion. Whether we need to worry about the difference between the equilibrium bond length and the experimentally determined average bond length and any uncertainty in these values depends on the purpose for which we are using it. In most of the discussions in this book we indicate whether the quoted value is an experimental or a calculated value, but do not differentiate between different experimental methods. We consider that the majority of the bond lengths we quote are accurate to within ± 1 pm and most of the bond angles to ± 2°. More detailed discussions of the differences between interatomic distances obtained by different methods have been given by Gillespie and Hargittai (1991) and by Ebsworth, Rankin, and Craddock (1987).

Bond lengths have usually been, and still often are, measured in angstroms (Å) but, with the advent of SI units, the nanometer (10^{-9} m) and the picometer (10^{-12} m) are now being used more frequently. In this book we express bond lengths and other molecular dimensions in picometers, which is for many purposes a more convenient unit than the angstrom (1 Å = 100 pm).

The length of the bond between two given atoms in predominately covalent molecules often varies only slightly from one molecule to another, although there are many exceptions to this generalization. If the exceptions are ignored, it is possible to divide the approximately constant length of a given type of bond into a contribution from each atom that is known as the **covalent radius** of the atom. Covalently radii are a useful property of an atom in a molecule because summing them for two atoms A and B gives an approximate value for the length of a covalent A—B bond. This radius is sometimes called the **atomic radius,** but the term "covalent radius" is to be preferred because it clearly refers to an atom forming a covalent bond in a molecule, not to the free atom. Table 2.1 gives values for the covalent radii for elements in groups 13–18. Values are not given for the elements in groups 1 and 2, which do not form any predominately covalent molecules, and they are not given for He, Ne, and Ar because these elements are not known to form any stable molecules.

The covalent radii for most of the elements were obtained by taking one-half of the length of a single bond between two identical atoms. For example, the covalent radius of sulfur is obtained from the length of the S—S bond in the S_8 molecule:

$$r(S) = \tfrac{1}{2}d(S—S) = \tfrac{1}{2} \times 208 \text{ pm} = 104 \text{ pm}$$

And the covalent radius of carbon can be obtained from the C—C bond length in diamond:

$$r(C) = \tfrac{1}{2}d(C—C) = \tfrac{1}{2} \times 154 = 77 \text{ pm}$$

For many molecules covalent radii are additive to within ±2 pm. For example,

$$d(C—S) = r(C) + r(S) = 77 + 104 \text{ pm} = 181 \text{ pm}$$

Table 2.1   Covalent Radii (pm) for the Elements in Groups 13–18

			Group		
13	*14*	*15*	*16*	*17*	*18*
				H	He
				37	—
B	C	N	O	F	Ne
88	77	70	65	60	—
Al	Si	P	S	Cl	Ar
143	117	110	104	99	—
Ga	Ge	As	Se	Br	Kr
125	122	121	117	114	111
In	Sn	Sb	Te	I	Xe
150	140	141	135	133	130

which compares well with the experimentally determined values of 180.7 pm in $S(CH_3)_2$ and 181.4 pm in $HSCH_3$.

There has been considerable uncertainty and disagreement concerning the values to be adopted for the covalent radii of O and F and to a lesser extent that of N because satisfactory values cannot be obtained by taking one-half of the N—N, O—O, and F—F bond lengths (Box 2.1). Fortunately this is not of great importance because oxygen and fluorine in particular form very few predominately covalent molecules. Because the hydrogen atom has only one electron and no inner core, its apparent radius in molecules is quite variable. The value of 37 pm given in Table 2.1 was obtained from the length of the bond in H_2, but in many molecules it has a radius of approximately 30 pm.

▲ BOX 2.1 ▼
The Covalent Radii of Nitrogen, Oxygen, and Fluorine

Two different sets of values for these radii have commonly been given in the past: those due to Schomaker and Stevenson (1941) and those due to Pauling (1960). These values together with those from Table 2.1 are given in Table Box 2.1. The Schomaker–Stevenson values were obtained from the lengths of the bonds in the N_2H_4, H_2O_2, and F_2 molecules as they were known at that time. The most recent values for the lengths of these bonds give only very slightly different values. However, it is widely recognized that the F—F bond in F_2, the O—O bond in H_2O_2, and the N—N bond in N_2H_4 are abnormally weak, as is shown by the following bond energies: F—F, 155; Cl—Cl, 240; O—O, 142; S—S 260; N—N, 167; P—P, 201 kJ mol^{-1}. So it is reasonable to conclude that these bonds are also abnormally long and that therefore the "normal" covalent radii of nitrogen, oxygen, and fluorine cannot be obtained from these bond lengths.

The values for the covalent radii of N and O given in the table do not differ significantly from the Pauling values, but the value for fluorine is a little smaller. They were obtained by extrapolation of the values for the other period 2 elements (Robinson et al., 1997). In any case the covalent radii of oxygen and fluorine are of little use because, as we shall see later, essentially all bonds formed by these elements, except the O—O, O—F, and F—F bonds, which are abnormally weak and long, have too great an ionic character to justify the use of covalent radii to calculate bond lengths.

Table Box 2.1 Values for the Covalent Radii of Nitrogen, Oxygen and Fluorine

Atom	Pauling	Schomaker and Stevenson	Table 2.1
N	70	74	70
O	66	74	65
F	64	72	60

The concept that the atoms of an element have a constant characteristic covalent radius is clearly only a rough approximation, inasmuch as we might expect that the radius of an atom would depend, to some extent, on the oxidation state of the element and on the number and nature of the attached atoms or groups that are conveniently called **ligands.** Another important limitation is that only homonuclear bonds are fully covalent. All bonds between different atoms are polar, their ionic character depending on the difference in the electronegativities of the bonded atoms. We discuss the effect of polarity on bond lengths in Section 2.5. It is common practice to deduce information about the nature of bonds from their lengths by comparing an observed bond length with that calculated by adding the covalent radii of the atoms forming the bond. Differences from the calculated values are then often interpreted in terms of *multiple-bond character (bond order)* or *polarity (ionic character).*

◆ 2.3   Multiple Bonds and Bond Order

The **order** *of a bond may be defined as the number of electron pairs that constitute the bond.* Thus the bond orders of single, double, and triple bonds are respectively 1, 2, and 3. As the number of electron pairs forming the bond increases, the attraction of the bonding electrons for the two atomic cores increases, so the bond strength increases and the bond length decreases.

A well-known example of the effect of bond order on bond length is provided by the bonds in ethane, ethene, and ethyne, which have the lengths of 154, 134, and 120 pm, respectively. Covalent radii for doubly and triply bonded atoms can be obtained from double and triple bond lengths in the same way as for single bonds. Some values are given in Table 2.2.

2.3.1   Resonance Structures

In many molecules the bonds between two given atoms have lengths that are intermediate between those of single and double bonds or between double and triple bonds. A familiar example is benzene for which the Lewis structure is

Table 2.2   Single, Double, and Triple Bond Radii (pm)

	C	N	O	P	S
Single bond	77	70	66	110	104
Double bond	67	61	57	100	94
Triple bond	60	55	52		

which implies that there are alternate single and double bonds. However, all six CC bonds have the same length of 140 pm, which is intermediate between that for a single bond and a double bond, C—C and C=C, respectively. Clearly the Lewis structure for benzene is inadequate. We can get a better description of the bonding in benzene if we assume that the structure of benzene is intermediate between the two possible Lewis structures

When Lewis structures are used in this way they are called **resonance structures.** The concept of resonance, which was introduced by Pauling, has a quantum mechanical basis, but it can be understood without going into its quantum mechanical basis at this point. The two resonance structures for benzene imply that six of the electrons—those forming the second component of each of the three double bonds—cannot be considered to be localized as pairs in three of the six CC bonding regions, as in either of the two resonance structures; rather, they are spread (delocalized) over all six bonding regions. Each bond can be thought of as consisting of one shared pair and one-half a shared pair, in other words, three electrons.

So each bond has a bond order of 1.5, which is consistent with the observed bond length. These two resonance structures are often called Kekulé structures because they were first proposed in 1865 by Kekulé, who imagined that the molecule converted very rapidly from one form to the other. This, however, is not the case: the molecule never has either of the Kekulé structures but only a single structure, which is intermediate between these two hypothetical structures and is approximately represented as follows:

where the circle represents the six delocalized electrons, and the H atoms and the CH bonds have been omitted in accordance with a common convention.

There are many other molecules in which some of the electrons are less localized than is implied by a single Lewis structure and can therefore be represented by two or more resonance structures. For example, the three bonds in the carbonate ion all have the same length of 131 pm, which is intermediate between that of the C—O single bond in methanol (143 pm) and that of the C=O double bond in methanal (acetaldehyde) (121 pm). So the carbonate ion can be conveniently represented by the following three resonance structures:

In these structures one pair of electrons on each oxygen is a lone pair in two of the structures and a bond pair in the other structure. Thus two of the electrons associated with each oxygen atom are not as localized as a single Lewis structure depicts. They behave partly as a lone pair and partly as a bonding pair.

> The need to use two or more resonance structures to describe the bonding in a molecule is a reflection of the inadequacy of Lewis structures for describing the bonding in molecules in which some of the electrons are not as localized as a Lewis structure implies.

Although there are better ways of describing delocalized electrons, because Lewis structures are simple, resonance structures are still often used in describing the bonding in a molecule. It is important to remember that despite its rather misleading name, resonance is not a phenomenon. Resonance structures are simply a crude way of representing an electron distribution that cannot be described by a single Lewis structure. Nevertheless, resonance is sometimes said to increase the stability of a molecule in the sense that the energy of a molecule is lower than that estimated for two or more equivalent resonance structures, such as the Kekulé structures of benzene, and the difference is called the **resonance energy.** However, the source of this extra stability is not a phenomenon associated with resonance but rather the decreased repulsion energy that arises from the electrons being farther apart (i.e., more delocalized), than in either of the resonance structures, so that resonance energy is also called **delocalization energy.**

> Resonance or delocalization energy is not a real energy inasmuch as it is not something that can be measured. It is simply the difference between the actual energy of the molecule and the energy of two or more hypothetical resonance structures.

Resonance structures have also been used to describe the polarity of bonds. For example, H—Cl can be described by the two resonance structures:

$$H—Cl \qquad H^+ \ Cl^-$$

The first represents a hypothetical HCl molecule with a purely covalent bond in which the two bonding electrons are equally shared between the two atoms, and the second a hypothetical molecule with a purely ionic bond in which both the bonding electrons have been transferred to the chlorine atom. In this case the two resonance structures do not necessarily contribute equally to the description of the real structure, but they may be given weights appropriate to describe the estimated or assumed polarity of the molecule. However, *it is not necessary to use resonance structures to describe a molecule such as H—Cl if we recognize that a bond between atoms of different electronegativity is necessarily polar.* To emphasize the polarity, the common convention of adding $\delta+$ and $\delta-$ signs to the appropriate atoms can be followed, as in $H^{\delta+}$—$Cl^{\delta-}$.

Multiple bonds between atoms with different electronegativities such as the C=O and S=O bonds, are also necessarily polar. This polarity is frequently represented by resonance structures such as

$$C\!=\!O \text{ and } C^+\!\!-\!O^- \qquad S\!=\!O \text{ and } S^+\!\!-\!O^-$$

where C=O represents a hypothetical purely covalent double bond and C^+—O^- represents a covalent bond plus an ionic bond. This description of a polar double bond to oxygen is somewhat limited in that it does not allow for the possibility that the oxygen might have a charge of greater than 1.0, as we shall see later is indeed the case for some SO and PO bonds. Thus the two resonance structures

$$X\!=\!O \qquad \text{and} \qquad X^{2+}O^{2-}$$

would give a more general description of a polar X=O bond. But the polarity can be equally well represented by $C^{\delta+}\!=\!O^{\delta-}$, although this representation is less commonly used for multiple bonds than for single bonds.

Before discussing the effect of bond polarity on bond lengths it will be convenient to introduce the concept of an ionic radius.

◆ 2.4 Ionic Radii

The concept of an **ionic radius** has proved useful for the discussion of the structures of ionic crystals, which in many, but by no means all cases can be understood in terms of the possible packing arrangements of ions of different sizes, assuming that ions can be represented as hard spheres with a constant characteristic radius. Because ionic bonds are necessarily between dissimilar atoms, the same technique used for obtaining covalent radii cannot be used to obtain ionic radii. The problem is to find a method for dividing the interionic distance into separate components for each ion. Once the radius of one ion has been obtained, the radius of any other ion can then be obtained, assuming that the radii are additive. It is possible to obtain some anion radii from the anion–anion distance in crystals containing large anions and small cations in which there is good reason to believe that the anions are close-packed and therefore touching each other, in which case one-half the distance between the anions can be taken as the anion radius.

The O^{2-} ion has often been chosen as the basis for deducing a set of ionic radii because it has the advantage of being found in combination with a wide range of elements. A widely used set of ionic radii was given by Pauling based on a radius of 140 pm for the oxide ion (Table 2.3). On the basis of these values, it is in principle possible to predict the interionic distance in any crystal and, assuming that the larger ions (usually the anions) pack as closely

Table 2.3 Pauling's Ionic Radii (pm)

Li^+	Be^{2+}	B^{3+}	C^{4+}			N^{3-}	O^{2-}	F^-
60	31	20	15			171	140	136
Na^+	Mg^{2+}	Al^{3+}	Si^{4+}	P^{5+}	S^{6+}	P^{3-}	S^{2-}	Cl^-
95	65	50	41	34	29	212	184	181
K^+	Ca^{2+}	Ga^{3+}	Ge^{4+}	As^{5+}	Se^{6+}	As^{3-}	Se^{2-}	Br^-
133	99	62	53	47	42	222	198	195
Rb^+	Sr^{2+}	In^{3+}	Sn^{4+}	Sb^{5+}	Te^{6+}	Sb^{3-}	Te^{2-}	I^-
148	113	81	71	62	56	245	221	216
Cs^+	Ba^{2+}	Tl^{3+}	Pb^{4+}	Bi^{5+}				
169	135	95	84	74				

as possible around the other ion (usually the cation), the geometric arrangement of the larger ions can also be predicted. For example, a ratio of the radius of the central atom A to that of the ligand X of 0.155 just allows 3 X ions to be packed around A in a trigonal planar arrangement. When the ratio $r_A/r_X = 0.225$, four X ions can just be packed around an A ion in a tetrahedral arrangement, and so on, as summarized in Table 2.4. Although there is a general increase in the coordination number of a cation with increasing ionic radius, there are many exceptions to predictions of ionic crystal structures made on the basis of radius ratios. For example, all the alkali metal halides except CsCl, CsBr, and CsI have the sodium chloride six-coordinated structure, even though radius ratios based on the radii in Table 2.4 indicate that only NaCl, NaBr, NaI, KBr, KI, and RbI should have this structure.

An important reason for the exceptions to the radius ratio predictions is that ions are not hard spheres but somewhat compressible, hence do not have a truly constant radius. Another reason for the inadequacy of the radius ratio rules, particularly when the anions are much larger than the cations, is that some structures are determined by the close packing of the anions, leaving the cations in "holes" between the anions. In such a case more anions may be packed around a cation of a given fixed radius than are predicted by the radius ratio, so that although the anions are touching each other, they are not touching the cation. However, if

Table 2.4 Radius Ratios and Coordination Polyhedra

Coordination Number	Minimum Radius Ratio	Coordination Polyhedron
3	0.155	Triangle
4	0.225	Tetrahedron
6	0.414	Octahedron
	0.528	Trigonal prism
7	0.592	Capped octahedron
8	0.645	Square antiprism
	0.668	Dodecahedron
	0.732	Cube
9	0.732	Tricapped trigonal prism
12	0.902	Icosahedron
	1.000	Cuboctahedron

Table 2.5   Shannon's Ionic Radii (pm)

Coord. No.	Li^+	Be^{2+}	B^{3+}	C^{4+}			N^{3-}	O^{2-}	F^-
2								135	128
3		17	1					136	130
4	59	27	11					138	131
6	76						146	140	133

Coord. No.	Na^+	Mg^{2+}	Al^{3+}	Si^{4+}	P^{5+}	S^{6+}	S^{2-}	Cl^-
4	99	48	39		31	26		
6	102	72	53		52	43	184	181
8	116							

Coord. No.	K^+	Ca^{2+}	Ga^{3+}	Ge^{4+}	As^{5+}	Se^{6+}	Se^{2-}	Br^-
4	137		47	39	34	28		
6	138	100	62	53	46	42	198	194
8	151	112						

Coord. No.	Rb^+	Sr^{2+}	In^{3+}	Sn^{4+}	Sb^{5+}	Te^{6+}	Te^{2-}	I^-
4			62	55		43		
6	152	110	80	69	60	56	221	220
8	161	126	92	81				

Coord. No.	Cs^+	Ba^{2+}	Tl^{3+}	Pb^{4+}	Bi^{5+}
4			75	65	
6	167	135	89	78	76
8	174	142	98	94	

we assume that the cation is always touching the anion, the radius for the cation cannot be constant but must increase with the coordination number. This is why Shannon has given a set of cation radii for different coordination numbers. A selection of these values is given in Table 2.5. Shannon obtained these values from the analysis of approximately 900 crystal structures of oxides and fluorides and the assumption of a radius for a six-coordinated oxide ion of 140 pm. These radii strictly apply only to fluorides and oxides, but to a reasonable approximation they are useful for compounds with other anions. Because cation radii increase with increasing coordination number of the ion, the sizes of cations should always be compared for the same coordination number.

Ionic radii are quoted in Tables 2.3 and 2.5 for a large number of cations including those of the elements in groups 13, 14, 15, and 16, which do not form predominately ionic bonds. These values were obtained by subtracting the fluoride or oxide ion radius obtained from predominantly ionic solids from the length of a bond that is not predominantly ionic. The very small values for the radii of "cations" obtained in this way do not bear much relation to the "real" size of the atom in the crystal or molecule.

A better idea of the "real" size of an ion in a molecule can now be obtained from a study of electron density distributions, which it has recently become possible to obtain from accurate X-ray crystallographic studies of crystals. Figure 2.2 shows a contour map of the electron density distribution obtained in an X-ray crystallographic study of crystalline sodium chloride. The position of minimum electron density between two adjacent ions seems to be

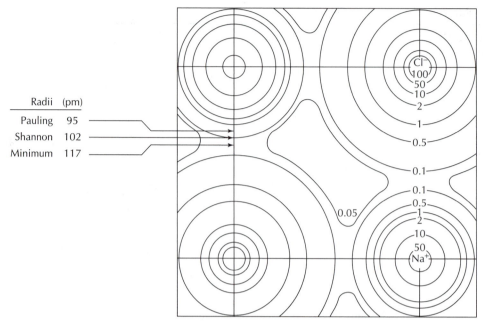

Figure 2.2 A contour plot of the electron density in a plane through the sodium chloride crystal. The contours are in units of 10^{-6} e pm^{-3}. Pauling shows the radius of the Na^+ ion from Table 2.3. Shannon shows the radius of the Na^+ ion from Table 2.5. The radius of the Na^+ ion given by the position of minimum density is 117 pm. The internuclear distance is 281 pm. (Modified with permission from G. Schoknecht, *Z. Naturforsch 12A*, 983, 1957 and J. E. Huheey, E. A. Keiter, and R. L. Keiter, *Inorganic Chemistry*, 4th ed., 1993, HarperCollins, New York.)

an obvious choice for identifying the boundary between two adjacent ions and thus for obtaining the radius of each ion. However, only a rather small number of crystal structures have been determined with the precision necessary to obtain an accurate electron density map. So ionic radii obtained in this way are not widely used at the present time. The radius of the sodium ion (117 pm) obtained from the electron density distribution in this way is significantly larger than either the Pauling or the Shannon radii while the radius of the chloride ion (164 pm) is significantly smaller. We must remember, though, that the ionic radius as defined by Pauling or Shannon is an arbitrary concept, which is useful provided we use a consistent set of values. The sum of the Pauling radii of Na^+ and Cl^- predicts an interionic distance of 276 pm, while the Shannon radii predict an interionic distance of 283 pm, values that are close to, but not exactly equal to, the observed distance of 281 pm.

The definition of the radius of an ion in a crystal as the distance along the bond to the point of minimum electron density is identical with the definition of the radius of an atom in a crystal or molecule that we discuss in the analysis of electron density distributions in Chapter 6. The radius defined in this way does not depend on any assumption about whether the bond is ionic or covalent and is therefore applicable to any atom in a molecule or crystal independently of the covalent or ionic nature of the bond, but it is not constant from one molecule or crystal to another. The almost perfectly circular form of the contours in Figure

2.2 shows that the ions have almost perfectly spherical charge clouds, indicating that the crystal is very nearly fully ionic, as we will discuss in detail in Chapter 6. In the next section we discuss how the lengths of bonds are affected by their polarity.

◆ 2.5   The Lengths of Polar Bonds

Having discussed the lengths of covalent and ionic bonds, and the concepts of covalent and ionic radii, we are in a position to discuss the lengths of polar bonds that are neither purely covalent nor purely ionic. It has long been recognized that a polar bond is shorter than calculated from the sum of the appropriate covalent radii. As was pointed out by Pauling, an X—Y bond has a bond energy that is generally greater than the average of the X—X and Y—Y bond energies (e.g., H_2, 436; Cl_2, 237; HCl, 431 kJ mol^{-1}). Pauling attributed this observation to the ionic character of the bond, that is, to the additional attraction between X and Y due to their charges. He used these energy differences as the basis of his electronegativity scale. Since a polar bond is stronger than expected for the corresponding hypothetical "pure" covalent bond, we expect that it would be shorter than predicted by the sum of the covalent radii, which should give the length of a "pure" covalent bond. Indeed many polar bonds, particularly X—F and X—O bonds, are shorter, and often much shorter, than the sum of the covalent radii. Schomaker and Stevenson (1940) proposed an empirical equation based on the difference in electronegativities of the two bonded atoms, which they claimed could be used to correct a bond length calculated from covalent radii. Their equation has the form

$$d_{AB} = r_A + r_B - k\,|\chi_A - \chi_B|$$

where d_{AB} is the predicted bond length, r_A and r_B are the covalent radii (pm), and $|\chi_A - \chi_B|$ is the absolute difference in the (Pauling) electronegativities of A and B. The radii for F, N, and O used in this equation are the Schomaker–Stevenson values (Box 2.1). Schomaker and Stevenson gave the constant k the value of 9 pm. Subsequently, to achieve a better fit with experimental data, Pauling modified this value to 8 pm for bonds involving one or two second-period elements and to 6, 4, or 2 pm for bonds formed by an element from periods 3, 4, and 5, respectively, with a more electronegative element. Other modifications of the equation have also been proposed, but they are all empirical and the equation is most often used in its original form.

   During the years following the proposal of the Schomaker–Stevenson equation, a large number of new bond lengths were accurately determined and many, particularly those to fluorine and oxygen, were found to be considerably shorter than the values predicted by the Schomaker–Stevenson equation. Some examples are given in Table 2.6. This equation is clearly not very useful for predicting bond lengths. One reason is that it is a purely empirical equation with no theoretical basis, being based on the rather small amount of experimental data available at the time it was proposed. Another important reason for the deficiency of the equation for X—O and X—F bonds in particular is that it is based on covalent radii obtained from the lengths of the bonds in H_2O_2 and F_2, which have long been recognized as being abnormally long and weak. Thus, an important reason for having to apply sig-

Table 2.6 Comparison of Observed Bond Lengths and Bond Lengths Calculated from the Sum of Covalent Radii

Molecule or Crystal	Bond	Bond Lengths (pm)		
		Observed	SS Equation[a]	Table 2.1[b]
SiO_2, $(SiH_3)_2O$	SiO	163	171	182
SiF_4	SiF	155	169	177
BF_3	BF	131	137	148
PF_3	PF	154	169	170
$SiCl_4$	SiCl	200	205	216

[a]Calculated from the Schomaker–Stevenson equation.

[b]Sum of the covalent radii (Table 2.1).

nificant corrections to predict heteronuclear bond lengths from the sum of covalent radii could simply be that the covalent radii that are used for F and O in particular are not valid. Indeed, there is no satisfactory way of obtaining covalent radii for these particular elements. This is not a serious problem, however, because the bonds between F or O and almost all other elements are predominantly ionic and so there is little use for covalent radii for these elements. In summary, it is clear that

> because of the additional electrostatic attraction between the charges on the atoms, polar bonds are stronger and shorter than the average of the two corresponding nonpolar covalent bonds.

◆ 2.6 Back-Bonding

As we have discussed, a correction to the length of a polar bond calculated from the sum of covalent radii has often been made using the Schomaker–Stevenson (SS) equation. But when it was found that in many cases even the corrected bond length was significantly longer than the experimental length (Table 2.6), rather than questioning the validity of the SS equation, or of the predominately covalent model on which it is based, Pauling suggested an additional effect. He proposed that since multiple bonds are shorter than single bonds, X—F and X—Cl bonds often have double-bond character arising from the donation of halogen lone pair electrons into the valence shell of the central atom A. This concept is usually called **back-bonding.** We consider two typical examples, BF_3 and SiF_4. The bond length in BF_3 is 130.7 pm. The bond length calculated from the SS equation is 137 pm (Table 2.6), and from the covalent radii in Table 2.1 it is 148 pm. To account for this discrepancy and because boron does not have a complete octet in BF_3, it was proposed that BF_3 be represented by the following three resonance structures:

in which there is donation of an electron pair from fluorine into the valence shell of boron, thus completing its octet and giving each bond 33% double-bond character or a bond order of $1^1/_3$. In these structures the boron has a formal negative charge and the fluorine a formal positive charge.

The SiF bond in SiF_4 has a length of 155.5 pm compared to the value of 169 pm calculated using the Schomaker–Stevenson equation and the value of 177 pm calculated from the sum of the covalent radii (Table 2.6). So Pauling wrote resonance structures such as the following:

$$
{}^+F{=}\overset{\displaystyle F}{\underset{\displaystyle F}{\overset{|}{\underset{|}{Si}}}}{}^-{-}F
\qquad
F{-}\overset{\displaystyle F}{\underset{\displaystyle F^+}{\overset{|}{\underset{\|}{Si}}}}{}^-{-}F
\qquad
F{-}\overset{\displaystyle F}{\underset{\displaystyle F}{\overset{|}{\underset{|}{Si}}}}{}^-{=}F^+
\qquad
F{-}\overset{\displaystyle F^+}{\underset{\displaystyle F}{\overset{\|}{\underset{|}{Si}}}}{-}F
$$

to account for the SiF bond length. If we add further resonance structures such as

$$
{}^+F{=}\overset{\displaystyle F^-}{\underset{\displaystyle F}{\overset{|}{\underset{|}{Si}}}}{-}F
\qquad\qquad
F{=}B{\Large\diagdown}\!\!\!\!\overset{\displaystyle F^-}{}_{\displaystyle F}
$$

in an attempt to describe the polarity of the bonds, and to conform to the octet rule in the case of SiF_4, the resonance description of these molecules becomes cumbersome and confusing. Morever, these descriptions of the bonding in BF_3 and SiF_4 in which fluorine has a formal positive charge are inconsistent with the large negative charge on fluorine that has been calculated for these molecules, as we discuss in Chapter 6. There is no reason to think that the lengths of these bonds cannot be adequately accounted for in terms of the large charges on their atoms. We make the following conclusion:

> The concept of back-bonding is not necessary to account for the lengths of polar bonds that are shorter than the sum of the covalent radii. These bonds are short because of the attraction between the atoms due to their opposite charges.

Back-bonding has usually been discussed in terms of the orbital model (Chapter 3), and we will revisit it again in later chapters. For the moment we need only emphasize that since the apparently short bond lengths can be accounted for in terms of the polarity of the bonds. Bond lengths do not provide any compelling evidence for the concept of back-bonding.

◆ 2.7 Bond Dissociation Energies and Bond Enthalpies

The energy needed to rupture a bond in a molecule is important not only for understanding the nature of a particular bond but also for understanding the reactivity of a molecule. However, this quantity cannot be obtained as directly and easily as bond lengths from experimental measurements.

We saw in Figure 2.1 that the energy needed to rupture the bond in a diatomic molecule, the **bond dissociation energy** is the energy ΔU_{el}. This is the energy that can be calculated by an ab initio calculation and refers to a molecule that has no vibrational, rotational, or translational energy. In other words, the molecule is at 0 K and has no zero-point vibrational energy. Experimentally we measure something slightly different, namely, ΔU_{298}, which refers to the dissociation of a molecule at 298 K under constant volume conditions. Because the majority of experimental measurements are carried out at constant pressure rather than at constant volume, the quantity that is generally more accessible is the enthalpy change ΔH, which includes $P\Delta V$ work and generally amounts to approximately 2.5 kJ mol^{-1}. For the dissociation of the hydrogen molecule at 298 K we have

$$\Delta H_{298} = \Delta U_{298} + P\,\Delta V = 433 + 2.5 \text{ kJ mol}^{-1} = 436 \text{ kJ mol}^{-1}$$

ΔH_{298} is called the **bond dissociation enthalpy** or simply the **bond enthalpy.** Bond enthalpies are often also called **bond energies** because the small difference between the two values (ca. 2.5 kJ mol^{-1}) can be ignored in many cases, particularly for polyatomic molecules.

For polyatomic molecules, the dissociation energy can be measured directly only for the weakest bond, and even then the value may only be approximate because the energies of the other bonds in the molecule generally change when one bond is broken. To obtain the energies of other bonds, some assumptions must be made. For molecules of the type AB$_n$ with only one type of bond, the enthalpy of atomization, that is, the enthalpy change for the reaction

$$AB_n(g) \rightarrow A(g) + nB(g)$$

is the energy needed to break all n A—B bonds so that $1/n$th of this energy is the **average bond enthalpy** for an AB bond. For example, the enthalpy of atomization of methane is 1663 kJ mol^{-1}, so the average bond enthalpy of the C—H bond in methane is 415.8 kJ mol^{-1}. However the energy needed to break just one CH bond is not equal to this value:

$$CH_4 \rightarrow {\cdot}CH_3 + H, \qquad \Delta H° = 439 \text{ kJ mol}^{-1}$$

There is a discrepancy of 23.2 kJ mol^{-1} between the two values because breaking one bond leads to a redistribution of the electrons in the remaining $\cdot CH_3$ molecule, so that the three remaining CH bonds are slightly different from the four original CH bonds. Of course the sum of the energies needed for this step and the following steps

$$\cdot CH_3 \rightarrow \cdot CH_2 + H$$

$$\cdot CH_2 \rightarrow \cdot CH + H$$

$$\cdot CH \rightarrow \cdot C\cdot + H$$

is equal to the enthalpy of atomization of methane, namely, 1663 kJ mol^{-1}.

When there are bonds of two or more kinds in a molecule, the determination of the bond enthalpies is slightly more complicated and is based on the assumption that bond enthalpies can be transferred from one molecule to another, at least to a reasonable approximation. For example, the enthalpy of atomization of ethane, which is 2826 kJ mol^{-1}, is the sum of six C—H bond enthalpies and one C—C bond enthalpy. The C—C bond enthalpy can be determined only if we make an assumption about the C—H bond enthalpy. If we make the rea-

Table 2.7 Mean Bond Enthalpies (kJ mol⁻¹): Homonuclear Bonds

H—H	B—B	C—C	N—N	O—O	F—F
436	301	348	159	138	155
		Si—Si	P—P	S—S	Cl—Cl
		196	197	266	237
		Ge—Ge	As—As	Se—Se	Br—Br
		163	177	193	190
		Sn—Sn	Sb—Sb	Te—Te	I—I
		152	142	126	140

sonable assumption that it is the same as in methane, we obtain a value of 331.3 kJ mol⁻¹ for the C—C bond enthalpy in ethane. Values of bond enthalpies obtained in this way differ somewhat from molecule to molecule, but the average of a number of such values gives a reasonably good approximation for the bond enthalpy for any given bond. Values of mean bond enthalpies are given in Tables 2.7 and 2.8.

Table 2.9 gives mean bond enthalpies for multiple bonds. These bond enthalpy values correlate well with bond order and bond length, increasing with increasing bond order, while bond lengths decrease with increasing bond order. For two given atoms, double bonds are invariably stronger—that is, they have higher bond enthalpies and are correspondingly shorter than single bonds. Triple bonds are even stronger and shorter. However, a carbon–carbon double bond is less than twice as strong as a single bond, and a triple bond is less than three times as strong. This also applies to phosphorus–phosphorus bonds and to sulfur–sulfur bonds. The two electron pairs in a double bond and the three pairs in a triple bond are closer together than if they were forming single bonds, and so there is an increased repulsion between them that results in the bond enthalpy being smaller than the sum of two or three single-bond enthalpies. However, this is not the case for NN and OO bonds, where a double bond is more than twice as strong as a single bond and a triple bond more than three times as

Table 2.8 Mean Bond Enthalpies (kJ mol⁻¹): Heteronuclear Bonds

	B	C	N	O	F	Si	P	S	Cl
X—H		413	389	463	565				
X—O	523	335	113	143	184	464	368		207
X—F	613	485	283	184	155	565	490	284	239
X—Cl	456	326	201	205	249	381	326	253	239

Table 2.9 Mean Bond Enthalpies (kJ mol⁻¹): Multiple Bonds

	CC	NN	OO	CN	CO
Single	348	159	138	293	335
Double	619	418	497	616	707
Triple	812	946		879	1070

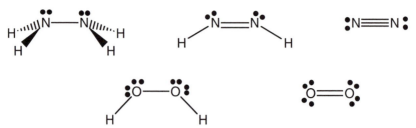

Figure 2.3 Lewis diagrams of some molecules of nitrogen, oxygen, and showing the increased separation of the lone pairs from the singly bonded, to the doubly bonded, and to the triply bonded molecules.

strong. The usual explanation for this unexpected order of bond strengths is based on the unexpected relative weakness of the N—N, O—O, and F—F single bonds, which has been attributed to strong repulsive interactions between the lone pairs in the crowded valence shells of these very small period 2 molecules. In the doubly bonded molecules, the lone pairs are further apart and their interaction is reduced. In the triply bonded molecules they are still further apart and their interaction is further reduced (Figure 2.3). The much larger size of period 3 atoms means that the magnitude of these repulsions is considerably reduced in their molecules. Thus, for example, P=P double and P≡P triple bonds, like C=C and C≡C bonds, have bond enthalpies that are less than twice the single-bond enthalpy and three times the single-bond enthalpy, respectively. However, this explanation needs to be further studied by a detailed analysis of the electron distribution in these molecules.

It is interesting to note that the strongly polar B—F and Si—F bonds have the largest known single-bond enthalpies, which must be a reflection of the nature of the bonds, as we will discuss in Chapters 8 and 9.

◆ 2.8 Force Constants

The force constant that is associated with the stretching vibration of a bond is often taken as a measure of the strength of the bond, although it is more correctly a measure of the curvature of the potential energy function around the minimum (Figure 2.1): that is, the rigidity of the bond. For a diatomic molecule, the frequency of vibration ν is determined by the force constant k and the reduced mass $\mu = m_1 m_2 / (m_1 + m_2)$, where m_1 and m_2 are the masses of the two atoms:

$$\nu = \frac{1}{2\pi} \left(\frac{k}{\mu} \right)^{1/2}$$

For polyatomic molecules, the stretching force constant for a particular bond cannot in general be obtained in an unambiguous manner because any given vibrational mode generally involves movements of more than two of the atoms, which prevent the expression of the observed frequency in terms of the force constant for just one bond. The vibrational modes of a polyatomic molecule can be analyzed by a method known a normal coordinate analysis to

Table 2.10 Vibrational Stretching Frequencies and Force Constants

Diatomic Molecules			Bonds		
	ν (cm^{-1})	k (10^2Nm^{-1})		ν (cm^{-1})	k (10^2Nm^{-1})
H_2	4395	5.70	≡C—H	2960	4.79
N_2	2360	22.98	⟩C—H	3300	5.85
O_2	1580	11.77	—C≡C—	2050	15.6
CO	2138	18.47	⟩C=C	1650	9.6
F_2	892	4.45	—C≡N	2100	17.7
Cl_2	546	3.19	⟩C=O	1700	12.1
Br_2	319	2.46	—O—H	3680	7.66
I_2	215	1.76	⟩C—Cl	650	3.64
HF	4138	9.66			
HCl	2991	5.16			
HBr	2650	4.12			
HI	2310	3.12			

give values of the force constants of individual bonds. However, when it is reasonable to assume that the stretching motion of one particular bond makes the main contribution to a particular normal mode, an approximate value for the stretching force constant of this bond can be obtained by treating the molecule as a diatomic. Some typical values for bond stretching force constants are given in Table 2.10. The values for bonds in polyatomic molecules are useful as an approximate measure of the bond strength or, more exactly, the resistance of the bond to stretching or compression. Because deep potential wells generally have steep sides, strong bonds with large dissociation energies generally have high force constants, although there is no general relationship between dissociation energies and force constants.

◆ 2.9 Dipole Moments

A point charge $+q$ separated from an equal and opposite point charge $-q$ by a distance d constitutes an **electric dipole.** An electric dipole has a **dipole moment μ,** which is a vector with magnitude $\mu = qd$, which is assumed to be acting in the direction from $+q$ to $-q$ (Figure 2.4).

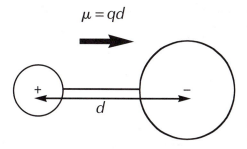

Figure 2.4 The dipole moment of a hypothetical purely ionic molecule with spherical ions.

The dipole moment of a molecule can be obtained from a measurement of the variation with temperature of the dielectric constant of a pure liquid or gaseous substance. In an electric field, as between the electrostatically charged plates of a capacitor, polar molecules tend to orient themselves, each one pointing its positive end toward the negative plate and its negative end toward the positive plate. This orientation of the molecules partially neutralizes the applied field and thus increases the capacity of the capacitor, an effect described by saying that the substance has a dielectric constant greater than unity (80 for liquid water at 20°C). The dipole moments of some simple molecules can also be determined very accurately by microwave spectroscopy.

A molecule has a dipole moment when the center of positive charge resulting from the nuclear charges does not coincide with the center of negative charge due to the electrons. Such molecules are called **polar molecules.** A homonuclear diatomic molecule, such as Cl_2, has a centrosymmetric distribution of electrons so that it has a zero dipole moment. A heteronuclear diatomic molecule in general has an unsymmetrical electron distribution because the different electronegativities of the two atoms result in a polar bond in which the atoms carry partial charges, as discussed in Chapter 1.

Values of the dipole moment of some diatomic molecules are given in Table 2.11. The SI unit of dipole moment is the coulomb-meter (C·m). This is a very large unit, so in Table 2.11 we use the unit 10^{-30} C·m. Dipole moments are often quoted in an older unit, the debye (D): 1 D = 3.24×10^{-30} C·m. We can see from Table 2.11 that the dipole moment of a diatomic molecule usually reflects the difference between the electronegativities of the two atoms.

Because the dipole moment of a diatomic molecule is qd, it would appear that if we knew the interatomic distance (bond length) d, we should be able to calculate the atomic charges $\pm q$. For example, the bond length of the HCl molecule is 127 pm and the dipole moment is 3.44×10^{-30} C·m, so we have

$$\mu = qd = 3.44 \times 10^{-30} \text{ C·m}$$

$$q = \frac{3.44 \times 10^{-30} \text{ C·m}}{127 \times 10^{-12} \text{ m}} \cdot \frac{1 \text{ electron}}{1.6 \times 10^{19} \text{ C}} = 0.17 \text{ electron}$$

$$\delta(\text{H}) = +0.17 \quad \delta(\text{Cl}) = -0.17$$

The simplicity of this argument has led to its being more widely used than is justified, and atomic charges obtained in this way have been used as a measure of the ionic character of

Table 2.11 Dipole Moments μ
(Cm · m × 10^{-30}) of Diatomic Molecules

Molecule	μ
HF	5.93
HCl	3.44
HBr	2.69
HI	1.46
Br$_2$	0
BrCl	1.42
BrF	1.68

a bond. For example, the HCl molecule is said to be 17% ionic or to have a 17% ionic character. In terms of two resonance structures

$$H—Cl \quad and \quad H^+ \; Cl^-$$

in which the first represents a pure covalent bond and the second a pure ionic bond, the ionic structure is said to contribute 17%. Similarly, since the calculated charges for the hydrogen fluoride molecule are ± 0.58, it has been said to have 58% ionic character. Although the charges calculated in this way are qualitatively consistent with the electronegativity differences between the atoms, they are not correct because the model on which they are based is oversimplified. The model assumes that the charge distribution around each nucleus is spherical and so behaves like a point charge situated at the position of the nucleus. However, in a real molecule the electronic charge cloud around an atom is distorted from a spherical shape so that the center of negative charge on an atom does not coincide with the nucleus, creating a small dipole called an **atomic dipole.** For example, if we form a diatomic molecule by placing a negative ion next to a positive ion, the positive ion will attract the electron cloud of the negative ion pulling it toward itself so that the center of negative charge no longer coincides with the nucleus, thus creating an atomic dipole (Figure 2.5b). The positive ion is said to **polarize** the negative ion. Similarly, the negative ion pushes the electron density of the cation away, thus polarizing the cation and creating another atomic dipole, which is generally considerably smaller than the atomic dipole created on the anion (Figure 2.5b). The measured dipole moment of a molecule is then the sum of the **charge transfer moment** and the two **atomic dipole moments** (Figure 2.5b) and so is smaller than the change transfer moment (Figure 2.5a).

Because atomic dipoles may be quite large, atomic charges calculated without taking them into account may be considerably in error. For example, the measured dipole moment of CO is only 0.37×10^{-30} C·m, and its bond length is 113 pm which, if we ignore the atomic dipoles gives the very small charges C(-0.020) and O($+0.020$) despite the large electronegativity difference between carbon and oxygen, and acting in the direction opposite to

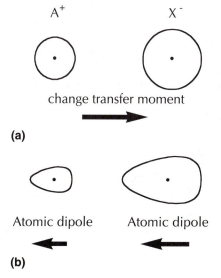

A^+ $\quad\quad X^-$

change transfer moment

(a)

Atomic dipole Atomic dipole

(b)

Figure 2.5 (a) Hypothetical ionic molecule with spherical ions and the corresponding charge transfer moment. (b) In a real molecule the ions are polarized leading atomic dipoles in each atom that oppose the charge transfer moment.

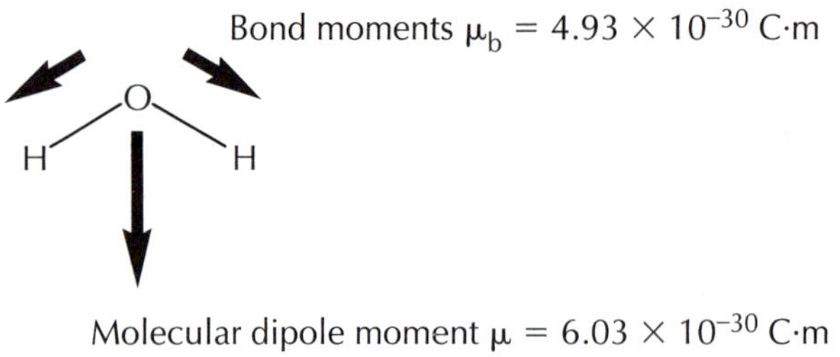

Bond moments $\mu_b = 4.93 \times 10^{-30}$ C·m

Molecular dipole moment $\mu = 6.03 \times 10^{-30}$ C·m

Figure 2.6 The dipole moment of the water molecule can be resolved into two OH bond moments. This does not give the true value of the bond moments because this procedure ignores the atomic dipoles.

that expected from the electronegativities. This is because the rather large atomic dipoles on carbon and oxygen oppose the ionic bond moment, as we will show in detail in Chapter 6. We see that the concept of ionic character based on atomic charges calculated from dipole moments is not soundly based. So the values for the ionic character of a molecule calculated from these charges, like the charges themselves, are also not reliable.

In a polyatomic molecule, the measured dipole moment can be regarded as the vector sum of a dipole moment due to each bond called the **bond moment.** For example, the dipole moment of the water molecule which is 6.03×10^{-30} C·m can be thought of as arising from two OH bond moments of magnitude 4.93×10^{-30} C·m making an angle of 104.5^- (the bond angle of the water molecule) with each other (Figure 2.6). Bond moments can be obtained in this way from the dipole moments of many molecules and average values for a few bonds are given in Table 2.12. Since, however, these listed values rest on the assumption that bond moments can be transferred from one molecule to another, which is only approximately true and may be considerably in error if there are large atomic dipoles, are only very approximate. Like many other bond properties (e.g., bond energies), bond moments are only approximate average values and are only approximately additive.

Nevertheless, the concept of a bond moment is useful even if precise values are uncertain. Because the dipole moment of a polyatomic molecule is the vector sum of the bond moments, high-symmetry molecules can have a zero dipole moment because even though they have very polar bonds, the vector sum of their bond moment is zero. In particular, we see in Figure 2.7a,b that each of the tetrahedral CCl_4 and the octahedral SF_6 molecule has a zero dipole moment. Of course if one of the ligands in a tetrahedral molecule is replaced by another ligand, as in $POCl_3$, the vector sum of the bond moments will then not in general be zero (Figure 2.7). This dependence of the dipole moment on molecular symmetry can be used to decide between possible molecular geometries. For example, bromine pentafluoride has a dipole moment of 4.92×10^{-30} C·m, which shows that it cannot have a trigonal bipyramidal geometry but is consistent with a square pyramidal geometry (Figure 2.8).

Table 2.12 Some Bond Dipole Moments

Bond	Moment (10^{-30}C · m)
O—H	−4.93
N—H	−4.34
C—F	4.89
C—Cl	5.15
C—O	2.78
C=O	7.8

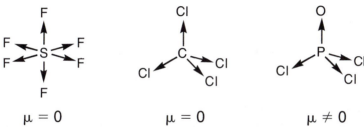

$$\mu = 0 \qquad\qquad \mu = 0 \qquad\qquad \mu \neq 0$$

Figure 2.7 The dipole moments of the symmetrical SF_6 and CCl_4 molecules are zero because the vector sum of the bond moments is zero. In contrast the dipole moment of $POCl_3$ is not zero because the molecule is less symmetrical and the vector sum of the three Cl bond moments is not equal and opposite to the CO bond moment.

Another example is provided by the *ortho, meta*, and *para* isomers of dichlorobenzene ($C_6H_4Cl_2$), which can be identified by their dipole moments, which decrease from ortho to meta to a zero value for the para isomer.

Ammonia (NH_3), has a dipole moment of 5.0×10^{-30} C·m, while nitrogen trifluoride (NF_3), in which the bonds are expected to be much more polar than in NH_3 owing to the larger difference in electronegativities, has a much smaller dipole moment of only 0.80×10^{-30} C·m. This very small dipole moment is at first sight somewhat surprising, but NF_3 is another example of a molecule in which there is a large atomic dipole. In particular, the lone pair on nitrogen gives a very asymmetric distribution of the electronic charge cloud around nitrogen, producing a large atomic dipole, often called a **lone pair dipole,** that opposes the NF bond dipoles, thus reducing the resultant dipole moment (Figure 2.9), whereas the atomic dipole in NH_3 reinforces the vector sum of the NH bond dipoles.

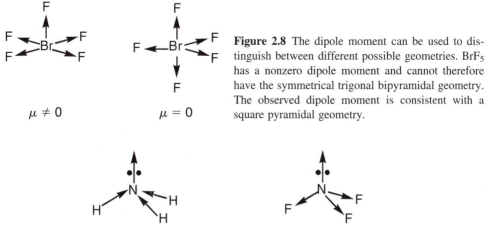

$$\mu \neq 0 \qquad\qquad \mu = 0$$

Figure 2.8 The dipole moment can be used to distinguish between different possible geometries. BrF_5 has a nonzero dipole moment and cannot therefore have the symmetrical trigonal bipyramidal geometry. The observed dipole moment is consistent with a square pyramidal geometry.

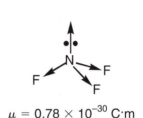

$$\mu = 4.76 \times 10^{-3} \text{ C·m} \qquad \mu = 0.78 \times 10^{-30} \text{ C·m}$$

Figure 2.9 The dipole moment of NH_3 is much larger than that of NF_3 because a large atomic dipole adds to the vector sum of the NH bond dipoles, whereas a large atomic dipole opposes the vector sum of the large NF atomic dipoles.

▶ References

E. A. V. Ebsworth, D. W. H. Rankin, and S. Cradock, *Structural Methods in Inorganic Chemistry,* 1987, Blackwell, Edinburgh.

R. J. Gillespie, and I. Hargittai, *The VSEPR Model of Molecular Geometry,* 1991, Allyn and Bacon, Boston.

R. J. Gillespie, and E. A. Robinson, *Inorg. Chem. 36,* 3022, 1997.

L. Pauling, *The Nature of the Chemical Bond.* 3rd Ed. 1960 Cornell University Press. Chapter 7 discusses bond lengths and covalent radii. Chapter 9 discusses ionic radii.

V. Schomaker, and D. P. Stevenson, J. Am. Chem. Soc. *63,* 37, (1941).

R. D. Shannon, Acta Crysallogr. *1325,* 925, 1976.

▶ Further Reading

J. E. Huheey, E. A. Keiter, and R. L. Keiter, *Inorganic Chemistry* 4th Ed. 1993, HarperCollins Chapter 6.

D. F. Shriver, P. W. Atkins, and C. H. Langford, *Inorganic Chemistry,* 1990, Freeman, New York, Chapter 2.

CHAPTER 3

SOME BASIC CONCEPTS
OF QUANTUM MECHANICS

■ ■ ■

◆ 3.1   Introduction

No real understanding of the chemical bond is possible in terms of classical mechanics because very small particles such as electrons do not obey the laws of classical mechanics. Their behavior is determined by quantum mechanics, which was developed in the second half of the 1920s. This development culminated in a mathematical formalism that we still use today. The interpretation of this formalism is still, however, a matter of debate, albeit among a small group of physicists and philosophers. Most chemists use quantum mechanics as a tool to obtain the wave function and corresponding energy and geometry of a molecule by solving the fundamental equation of quantum mechanics, called the Schrödinger equation. This equation cannot be solved exactly for any atom or molecule other than a hydrogen-like atom, so approximation methods must be used. These can produce solutions of great accuracy given sufficient time and computer power, but obtaining more accurate solutions more quickly at a reasonable cost remains a major challenge of present-day quantum chemistry.

In this chapter we give a brief review of some of the basic concepts of quantum mechanics with emphasis on salient points of this theory relevant to the central theme of the book. We focus particularly on the electron density because it is the basis of the theory of atoms in molecules (AIM), which is discussed in Chapter 6. The Pauli exclusion principle is also given special attention in view of its role in the VSEPR and LCP models (Chapters 4 and 5). We first revisit the perhaps most characteristic feature of quantum mechanics, which differentiates it from classical mechanics: its probabilistic character. For that purpose we go back to the origins of quantum mechanics, a theory that has its roots in attempts to explain the nature of light and its interactions with atoms and molecules. References to more complete and more advanced treatments of quantum mechanics are given at the end of the chapter.

◆ 3.2 Light, Quantization, and Probability

The debate about whether light should be described as a stream of particles or as an (electromagnetic) wave is older than quantum mechanics. Although Newton proposed that light consists of particles, Huygens produced arguments that favored a wave theory. Over the following two centuries an overwhelming amount of evidence supporting the wave theory was accumulated, and the debate culminated in Maxwell's complete quantitative wave theory of all types of electromagnetic radiation.

However, Planck's explanation of blackbody radiation and Einstein's explanation of the photoelectric effect reopened the debate because these explanations invoked the idea of quantization. It was postulated that light is not just emitted or absorbed in light quanta but that it travels through space as small bundles of energy called **photons.** Although photons are regarded as (massless) particles, their energy is remarkably expressed in terms of the frequency of a wave. Indeed, the energy of the photon, is given by $E = h\nu$, where $h = 6.626 \times 10^{-34}$ J·s is a constant called the Planck constant, and ν is the frequency of the light. Even though a photon has no mass, it has a momentum, which depends on the wavelength λ of the light and is given by $p = h/\lambda$. In summary, the energy of light is transmitted in the form of particle-like photons, which, however, have an energy that depends on the frequency of the light.

That light has a dual nature and behaves either like a wave or like a stream of particle-like photons is a fact we must accept, although it is nonintuitive. But remember, we have no direct experience of the behavior of very small particles such as electrons. Which model we use depends on the observations we are making. The wave model is appropriate when we are considering diffraction and interference experiments, but the particle (photon) model is essential when we are considering the interaction of light with individual atoms or molecules.

We now discuss a simple experiment that vividly illustrates the dual nature of light and introduces the concept of probability. In Young's slit experiment (Figure 3.1a), light from a single source is passed through two closely spaced slits. An interference pattern of light and dark lines (Figure 3.1b) is observed on a screen placed in the path of the light coming from the two slits. According to the wave theory, the interference pattern is produced by the interference of the waves arriving from the two slits. At certain points the waves reinforce each other, giving a wave of increased amplitude at other points the waves cancel each other, giving a resultant wave of decreased or zero amplitude.

Now, what happens if we use light of a *very weak intensity* and replace the screen by a bank of detectors such as photomultipliers, each of which can detect the arrival of a single photon? The photons are detected in what at first appears to be a random manner (Figure 3.2a,b). But if we observe the arrival of a sufficiently large number of photons at the bank of detectors, the interference pattern begins to emerge (Figure 3.2c,d). A large number of photons arrive at points corresponding to the bright lines—the maxima in the interference pattern—while very few photons arrive at points corresponding to the dark regions between the bright lines—the minima in the diffraction pattern (Figure 3.2e). We cannot say where any individual photon will be detected, only that it has a high probability of being detected at certain points. There is zero probability that the photon will be detected at the minima of the diffraction pattern. We see that the wave theory describes the *statistical* result of observing a large number of photons. The behavior of any individual photon cannot be pre-

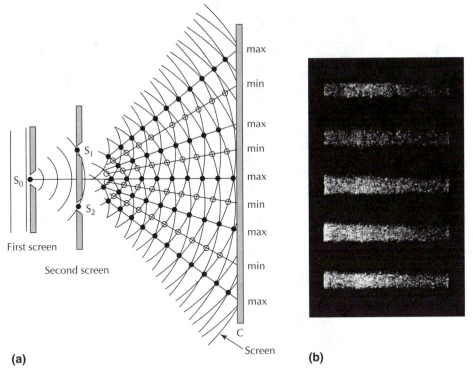

(a)   **(b)**

Figure 3.1 (a) Schematic diagram (not to scale) of Young's double-slit experiment. The narrow slits acts as wave sources. Slits S_1 and S_2 behave as coherent sources that produce an interference pattern on screen C. (b) The fringe pattern formed on screen C could look like this. (Reproduced with permission from R. A. Serway *Physics for Scientists and Engineers with Modern Physics,* 3rd ed, 1990, Saunders, Figure 37.1.)

dicted, but we can predict the *probability* of detecting a single photon at a particular point. This probability is proportional to the intensity of the light, that is, to the square of the amplitude of the wave, at that point. This is a very important conclusion because it will prove to be to valid for non-zero-mass particles as well, such as electrons.

◆ 3.3   The Early Quantum Model of the Atom

In 1913 Bohr proposed a model of the hydrogen atom to account for its spectrum. The spectrum of the hydrogen atom is called a *line spectrum* because it is not continuous but contains only a few well-defined wavelengths. This line spectrum shows that the energy of the electron in a hydrogen atom can have only certain definite energies—it is said to be *quantized.* Bohr's model was a combination of Rutherford's model of the atom with ideas from Planck's quantum theory, Einstein's photon theory of light, and some revolutionary postu-

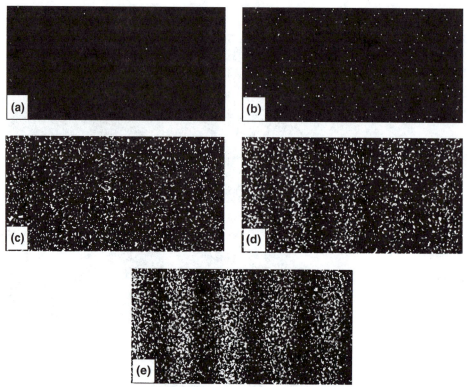

Figure 3.2 The cumulative pattern generated by photons sent one by one through a two-slit interferometer. Number of photons: (a) 10, (b) 100, (c) 3000, (d) 20,000, and (e) 70,000. (Reproduced with permission from Tonomura et al, *Amer. J. Phys. 57*, 117, 1987.)

lates that could not be justified within the framework of classical mechanics. Bohr proposed that the single electron in the hydrogen atom revolves around the nucleus in a circular orbit. According to Coulomb's law, the energy of the electron depends on its distance from the nucleus. And so the larger the orbit, the greater the energy of the electron. According to classical mechanics, the electron can have any orbit, and therefore any energy. To explain the line spectrum of hydrogen Bohr had to assume that only certain orbits and therefore certain energies are allowed.

By means of his model Bohr was able to account quantitatively for the spectrum of the hydrogen atom. But he could not explain *why* only certain orbits are allowed, nor could he account for the electron's continual rotation around the nucleus with constant energy because, according to electromagnetic theory, an electron rotating around a positive nucleus should emit electromagnetic radiation, continuously losing energy and spiraling into the nucleus. Moreover, attempts to expand the model to account for the spectra of atoms with more than one electron were not successful. The Bohr theory of the hydrogen atom specified the precise path of the electron, which is why it cannot be correct because, as we will see in the next section, the concept of a well-defined orbit for an electron is not consistent with quantum mechanics.

◆ 3.4 The Wave Nature of Matter and the Uncertainty Principle

In 1923 de Broglie made the bold suggestion that matter, like light, has a dual nature in that it sometimes behaves like particles and sometimes like waves. He suggested that material (i.e., non-zero-rest mass) particles with a momentum $p = mv$ should have wave properties and a corresponding wavelength given by

$$\lambda = \frac{h}{p} = \frac{h}{mv} \tag{3.1}$$

There was no experimental evidence for the wave nature of matter until 1927, when evidence was provided by two independent experiments. Davisson found that a diffraction pattern was obtained if electrons were scattered from a nickel surface, and Thomson found that when a beam of electrons is passed through a thin gold foil, the diffraction pattern obtained is very similar to that produced by a beam of X-rays when it passes through a metal foil.

In classical mechanics both the position of a particle and its velocity at any given instant can be determined with as much accuracy as the experimental procedure allows. However, in 1927 Heisenberg introduced the idea that the wave nature of matter sets limits to the accuracy with which these properties can be measured simultaneously for a very small particle such as an electron. He showed that Δx, the product of the uncertainty in the measurement of the position x, and Δp, the uncertainty in the measurement of the momentum p, can never be smaller than $h/2\pi$:

$$\Delta x \Delta p \geq \frac{h}{2\pi} \tag{3.2}$$

This relationship is called the **uncertainty principle** (Box 3.1).

If we increase the accuracy with which the position of the electron is determined by decreasing the wavelength of the light that is used to observe the electron, then the photon has a greater momentum, since $p = h/\lambda$. The photon can then transfer a larger amount of momentum to the electron, and so the uncertainty in the momentum of the electron increases. Thus any reduction in the uncertainty in the position of the electron is accompanied by an increase in the uncertainty in the momentum of the electron, in accordance with the uncertainty principle relationship. We may summarize by saying that there is no way of accurately measuring simultaneously both the position and velocity of an electron; the more closely we attempt to measure its position, the more we disturb its motion and the less accurately therefore we are able to define its velocity.

◆ 3.5 The Schrödinger Equation and the Wave Function

The picture provided by de Broglie inspired Schrödinger to propose in 1926 a powerful equation called the **Schrödinger wave equation,** which describes the behavior of very small particles such as electrons. From this equation we can obtain the form of the wave, called the wave function, and the corresponding energy E for any atomic or molecular system. Although some textbooks give plausible derivations for the Schrödinger wave equation, it ac-

▲ BOX 3.1 ▼
Applying the Uncertainty Principle

The uncertainty principle is of importance only for very small particles such as single electrons. This can be seen by comparing the accuracy with which we could determine the position of a very small *macroscopic* object with the accuracy of the determination of the position of an even smaller electron. Consider a dust particle with a mass of $1 \mu g = 10^{-9}$ kg, traveling in the wind with a speed of 10 m s^{-1}. If the speed of the particle is known with an accuracy of 0.1%, then the uncertainty in its momentum, Δp, is $0.1 \times 10^{-2} \times 10^{-9} \times 10$ m s$^{-1} = 1 \times 10^{-1}$ kg m s^{-1}. According to the uncertainty principle, $\Delta x \, \Delta p \geq h/2\pi$, the uncertainty in the dust particle's position, Δx, is at least $(h/2\pi)/\Delta p = 1 \times 10^{-34}$ kg m^2 s^{-1}/1×10^{-11} kg m s$^{-1} = 1 \times 10^{-23}$ m, which is approximately a hundred million times smaller than the dimensions of a nucleus. In contrast, consider an electron, which has a mass of only 9×10^{-31} kg, moving at the same speed as the dust particle (which is very slow for an electron). If we know its velocity to the same accuracy, then a similar calculation shows that the corresponding uncertainty in its position is at least 0.1 m. This uncertainty is about a *billion times* larger than typical atomic dimensions.

tually cannot be derived from anything else. It must be accepted as a fundamental postulate of quantum mechanics, just as Newton's equations are the fundamental postulates of classical mechanics. The justification for the Schrödinger equation, as with Newton's equations, is that it gives results in agreement with experiment.

A compact and completely general form of the Schrödinger equation is as follows

$$H\psi = E\psi \tag{3.3}$$

where ψ is the wave function, E is the energy, and H is an operator called the Hamiltonian operator. The differential operators d/dx and d^2/dx^2 are familiar examples of operators. The name Hamiltonian comes from Hamilton's equations of classical mechanics, which employ an analogous function to generalize Newton's laws of motion. The form of the Hamiltonian depends on the system under consideration. For an atom, the Hamiltonian contains a kinetic energy term and a potential energy term that results from the Coulomb attraction between the electrons and the nucleus and interelectron repulsion, and is given by the classical expression for the potential energy.

By solving this equation for any particular system, we can obtain the wave function and the corresponding energy E. In fact, an equation of this type in which an operator operates on a function to give the function times a constant has many solutions. Each solution is called a **state,** and the lowest energy state is called the ground state. Only certain of the possible solutions to this equation are acceptable, namely, those for which ψ is a continuous single-valued function. As we shall see shortly, ψ^2 represents a probability density that, to be physically reasonable, must change in a continuous way and must have only one value at each point in the system.

Thus in principle we can obtain the wave function and the energy for any atomic or molecular system, but insuperable mathematical difficulties prevent us from obtaining exact solutions to all but the very simplest one-electron systems. However, we can obtain approximate solutions, which are becoming increasingly easy to obtain and more accurate as modern computers gain in speed and power. To illustrate the use of the Schrödinger equation, we consider the simple case of a single particle moving in one dimension. For this system, the Hamiltonian operator H is

$$H = \left(-\frac{h^2}{8\pi^2 m} \cdot \frac{d^2}{dx^2} + V\right) \tag{3.4}$$

so that the Schrödinger equation for this system is

$$\left(-\frac{h^2}{8\pi^2 m} \cdot \frac{d^2}{dx^2} + V\right)\psi(x) = E\psi(x) \tag{3.5}$$

In this equation h is the Planck constant, m is the mass of the particle, and V is its potential energy. Since V is constant for a particle with no forces acting on it, V may be omitted from the equation, which simply means that the total energy E is only the kinetic energy. If we define the constant

$$k^2 = \frac{8\pi^2 m E}{h^2} \tag{3.6}$$

the Schrödinger equation 3.5 may be rewritten as follows, as a linear second-order differential equation:

$$\frac{d^2\psi}{dx^2} + k^2 = 0 \tag{3.7}$$

A convenient form of the general solution of Equation (3.7) is

$$\psi = A \cos kx + B \sin kx \tag{3.8}$$

where A and B are constants. Since there are no constrains or restrictions on k, it may have any value, and so the energy E may have any positive value. This means that the energy of a free particle is not quantized.

We now consider what happens when the particle is no longer free but is confined to a limited region of space. We do this by assuming that $V = 0$ for the region $x = 0$ to $x = l$ but everywhere outside this region $V = \infty$. Although this situation is physically unrealizable, it is a mathematically simple and useful model known as the **particle in a box** (Figure 3.3). It provides an illustration of some important quantum mechanical concepts without obscuring the principles with mathematical details.

The only solution for Equation (3.5) when $V = 0$ is $\psi = 0$. So that if ψ is to be single-valued and continuous, it must be zero at the walls, that is, at $x = 0$ and $x = l$. Thus the potential energy walls impose what are called *boundary conditions* on the form of the wave function. Figure 3.3 shows (a) the particle-in-a-box potential, (b) a wave function, that satisfies the boundary conditions and, (c) one that does not. We see that only certain wave func-

(a)

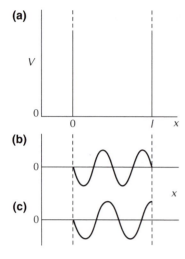

(b)

(c)

Figure 3.3 (a) The potential energy function assumed in the particle-in-a-one-dimensional-box model. (b) A wave function satisfying the boundary conditions. (c) An unacceptable wave function. (Reproduced with permission from P. A. Cox, *Introduction to Quantum Theory and Atomic Structure*, 1996, Oxford University Press, Oxford, Figure 2.6.)

tions will satisfy the boundary conditions, and since each is associated with a particular energy, we see that the energy of the system is quantized. The possible values of the energy follow from the boundary conditions and are given by the expression

$$E = \frac{n^2 h^2}{8ml^2} \tag{3.9}$$

where n, which can have any integral value greater than zero, is called a quantum number. Substituting the allowed energies in Equation (3.4) gives

$$\psi = B \sin(n\pi x/l) \tag{3.10}$$

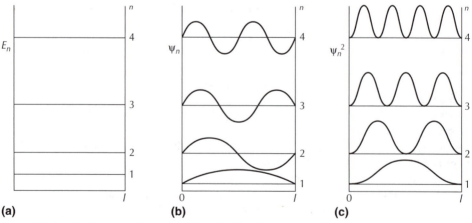

(a) **(b)** **(c)**

Figure 3.4 Particle-in-a-one-dimensional box. (a) The four lowest allowed energy levels ($n = 1, 2, 3$ and 4). (b) The corresponding wave functions ψ_n. (c) Probability densities ψ_n^2.

The four lowest energy wave functions are shown in Figure 3.4. The energy of the particle in its lowest energy state is called its **zero-point energy.** A wave has certain points at which it has a zero value. These points are called **nodes.**

The wave functions of the particle in a box are analogous to the waveforms for the vibration of the string that is fixed at both ends in a musical instrument. The string can vibrate to give only the fundamental or lowest energy note and the successive higher harmonics, which have an increasing number of nodes. In atomic and molecular systems, the energy is similarly quantized. Only certain wave functions and the corresponding energies are allowed because the electrons are confined to a certain region of space, like the violin string, as a consequence of the boundary conditions, as we saw in the example of a particle in a box.

◆ 3.6   The Meaning of the Wave Function: Probability and Electron Density

What is the meaning of the wave function? How should it be interpreted ? How can a particle also behave like a wave? These are questions that immediately come to mind.

Schrödinger was very much led by a comparison with real macroscopic waves, and he thought that an electron or other particle actually spread out in space. However, difficulties arose in trying to interpret a many-electron wave function in this way. Even for a single electron confined in a box, we can make the box as large as we like and it seems unreasonable to suppose that a single particle such as an electron can be spread out over a very large region of space. Every time we detect an electron, it behaves as if it has the same mass and charge, located in a very small region of space. An elegant solution to this dilemma was proposed by Born, who linked a wave function to a probability density.

We prepared the ground for the Born interpretation of the wave function by looking at the nature of light in Section 3.2, where we concluded that the behavior of photons is *statistical.* Since de Broglie showed that photons and electrons do not differ fundamentally from the standpoint of quantum mechanics, we now surmise that a *probabilistic* interpretation of the behavior of electrons may be successful as well. To sharpen the analogy, we argue as follows. The probability of detecting a photon at a given point is proportional to the intensity of the light at that point, which is proportional to the square of the amplitude of the light wave. So the probability of finding particle at a given point must equally be proportional to the square of the amplitude of a matter wave, that is to ψ^2. There are some systems for which ψ is a complex quantity. For these cases we must replace ψ^2 by $\psi\psi^*$, where ψ^* is the complex conjugate to ψ.

It does not make much sense to talk about the probability of finding a particle at a certain point x, y, z in space because a point is infinitely small, and there are an infinite number of points even in an extremely small volume, so that the probability of finding a particle at any one point must be zero. Instead we refer to the function $P(x, y, z) = \psi^2$ as the **probability density.** $P(x, y, z)dx\, dy\, dz\ (= P\, d\tau)$ is then the probability of finding the particle in the small element of volume $dx\, dy\, dz\ (= d\tau)$. Since the particle must be somewhere, we write

$$\int P\, d\tau = \int \psi^2 d\tau = 1 \tag{3.11}$$

But what does this probability density mean in physical terms? The electron should not be thought of as intrinsically smeared out as originally conceived by Schrödinger. In the Born

interpretation, an electron is either here or there, as if it were a sharply defined point particle. But, there is no way of exactly predicting where the electron will be at a given instant—
its behavior is intrinsically probabilistic. Suppose that at a particular moment we were able
to determine exactly where the electron was and to mark its position with a minute dot in
space. We repeat the observation a very large number of times, and each time put a dot in
the appropriate place. What we obtained would look like a "cloud," with the greatest density of dots indicating where there is the greatest probability of finding the electron. Another
analogy is to imagine that we could take a time exposure photograph of the electron as it
moves around the atom. We would obtain a blurred picture of the electron that was bright-
est where the electron is most likely to be found and less bright where it is less likely to be
found—the electron would appear to be spread out in a cloud.

If we multiply the probability density $P(x, y, z)$ by the number of electrons N, then we
obtain the **electron density distribution** or **electron distribution,** which is denoted by $\rho(x,$
$y, z)$, which is the probability of finding an electron in an element of volume $d\tau$. When in-
tegrated over all space, $\rho(x, y, z)$ gives the total number of electrons in the system, as expected. The real importance of the concept of an electron density is clear when we consider
that the wave function ψ has no physical meaning and cannot be measured experimentally.
This is particularly true for a system with N electrons. The wave function of such a system
is a function of $3N$ spatial coordinates. In other words, it is a multidimensional function and
as such does not exist in real three-dimensional space. On the other hand, the electron density of any atom or molecule is a measurable function that has a clear interpretation and exists in real space.

◆ 3.7 The Hydrogen Atom and Atomic Orbitals

The simplest atomic system that we can consider is the hydrogen atom. To obtain the Hamiltonian operator for this three-dimensional system, we must replace the operator d^2/dx^2 by
the partial differential operator

$$\nabla^2 = \frac{\partial^2}{\partial x^2} + \frac{\partial^2}{\partial y^2} + \frac{\partial^2}{dz^2}$$

The potential energy of the electron is given by $V = e^2/(x^2 + y^2 + z^2)^{1/2}$, where $(x^2 + y^2 + z^2)^{1/2}$
is the radial distance from the nucleus. So the Schrödinger equation for the hydrogen atom is

$$\left(\frac{-h^2}{8\pi^2 m} \nabla^2 - \frac{e^2}{(a^2 + y^2 + z^2)^{1/2}} \right) \psi(x, y, z) = E\psi \tag{3.12}$$

In this form the equation is rather cumbersome and not easily solved, so it is customary to
express it in spherical polar coordinates r, θ, and, ϕ, where r is the distance from the nucleus and θ and ϕ are angular coordinates, rather than in the Cartesian coordinates x, y, and
z. The relationship of the polar coordinates to the Cartesian coordinates is shown in Figure
3.5. In this form $V = e^2/r$, and the equation is easier to solve particularly because it can be
expressed as the product $R(r)\Theta(\theta)\Phi(\phi)$ of the three one-dimensional functions: R, the radial
function, and Θ and Φ, the angular functions. Corresponding to these three functions there
are three quantum numbers, designated n, l, and m.

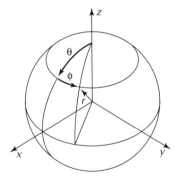

Figure 3.5 The relationship between polar coordinates and Cartesian coordinates.

The solutions to the Schrödinger equation for the hydrogen atom, which are called **orbitals,** are given in all standard textbooks of quantum mechanics. Rather than repeating the equations for these solutions, we illustrate them by means of diagrams. However, it is not quite simple to depict these orbitals, because to do so completely would require four coordinates, one for each of the three spatial coordinates and one for the wave function ψ, just as we need two dimensions to depict the solutions of a one-dimensional wave function (Figure 3.3). But as we will see, we can easily illustrate the separate radial and angular functions. The energy states of the hydrogen atom are given by Equation (3.13)

$$E_n = \frac{me^4}{8\varepsilon_0 h^2 n^2} \quad \text{where m = mass of the electron and } \varepsilon_0 = \text{permittivity of a vaccum} \quad (3.13)$$

which is also the expression found experimentally and deduced by Bohr, as we would expect if the equation gives a valid description of the atom. The energy depends only on the value of the quantum number n. The radial function $R(r)$ shows how the value of R varies along any radius from the nucleus. The radial functions for the $n = 1$, $n = 2$, and $n = 3$ states are shown in Figure 3.6. There are two radial functions for the $n = 2$ state corresponding to the two possible values of 0 and 1 for the quantum number l, which may take all integral values up to $n - 1$. The radial function for the $l = 0$ orbital has a maximum at the nucleus and node at a point along the radius, whereas the $l = 2$ orbital has a node at the nucleus. Orbitals for which $l = 0$ are called s orbitals, those for which $l = 1$ are called p orbitals. and those for which $l = 3$ are called d orbitals.

For an s orbital the angular part of the wave function is constant

$$(\Theta\Phi)_s = \text{constant} \quad (3.14)$$

which means that a plot of this function in a plane through the z axis is a circle and in three-dimensions is a sphere (Figure 3.7a). There are three p orbitals corresponding to the value of the third quantum number m, which may take the values . . . $l - 2, l - 1, 0, 1 + l, l + 2,$ . . . , which for $l = 1$ are the three values $m = -1, 0,$ and -1. These three orbitals are distinguished as p_x, p_y, and p_z. For the p_z orbital, the angular wave function $\Theta\Phi$ varies as cos θ:

$$(\Theta\Phi)_p = \text{constant (cos } \theta) \quad (3.15)$$

A plot of the angular function of $\Theta\Phi$ against the value of θ in a plane through the z axis has the form of a plot of cos θ and consists of two circles with a node in a line along the x axis

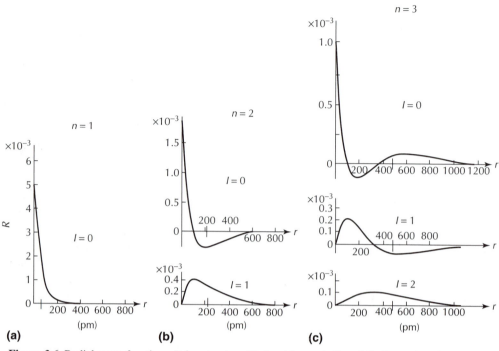

Figure 3.6 Radial wave functions R for atomic orbitals with $n = 1$, 2, and 3. (Reproduced with permission from J. E. Huheey, E. A. Keiter, and R. L. Keiter, *Inorganic Chemistry,* 4th ed., Harper Collins, 1993, Fig. 2.2; and G. Herzberg, *Atomic Spectra and Atomic Structure,* 1944, Dover, New York, 1944.)

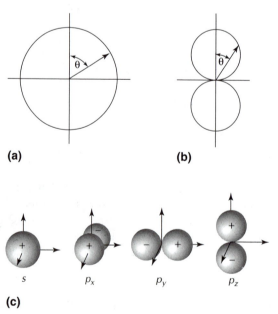

Figure 3.7 (a) Plot of the angular function $(\Theta\Phi)$ = constant for an s orbital in plane through the z axis. (b) Plot of $(\Theta\Phi)$ = constant (cos θ) for a p orbital in a plane through the z axis. (c) Three-dimensional plots of the angular function $(\Theta\Phi)$ for the s and p orbitals. (Adapted with permission from P. A. Cox, *Introduction to Quantum Theory and Atomic Structure,* 1996, Oxford University Press, Oxford, Figure 4.4.)

(Figure 3.7b) and in three dimensions of two spheres with a node in the xy plane (Figure 3.7c). The plots for the $2p_x$ and $2p_y$ orbitals are the same except that they are oriented along the x and y axes, respectively (Figure 3.7c).

To obtain pictures of the orbital $\psi = R\Theta\Phi$, we would need to combine a plot of R with that of $\Theta\Phi$, which requires a fourth dimension. There are two common ways to overcome this problem. One is to plot contour values of ψ for a plane through the three-dimensional distribution as shown in Figures 3.8a,c another is to plot the surface of one particular contour in three dimensions, as shown in Figures 3.8b,d. The shapes of these surfaces are referred to as the shape of the orbital. However, plots of the angular function $\Theta\Phi$ (Figure 3.7) are often used to describe the shape of the orbital $\psi = R\Theta\Phi$ because they are simple to draw. This is satisfactory for s orbitals, which have a spherical shape, but it is only a rough approximation to the true shape of p orbitals, which do not consist of two spheres but rather two squashed spheres or "doughnut" shapes.

Often we are more interested in the electron density distribution $\psi^2 = R^2\Theta^2\Phi^2$ than in ψ. Plots of R^2 the electron density distribution along a radius for the two $n = 2$ states, are shown in Figure 3.9a,b. Plots of $(\Theta\Phi)^2 = $ constant $\cos^2\theta$ (Figure 3.9c) show how the electron density varies with the angle θ. They are similar to the plots for $\Theta\Phi$ except the plots for the 2p orbitals are more elongated. The result of combining these two plots to obtain a plot of ψ^2 can be shown as a contour plot as Figure 3.10b,e. For an s orbital the plot of ψ^2 is spherical like the plot of ψ but for a p orbital it has a shape rather different from that of the angular part of ψ, which may be described as two squashed spheres or doughnut shapes separated by a plane in which ψ^2 is zero. We can also show the three-dimensional surface

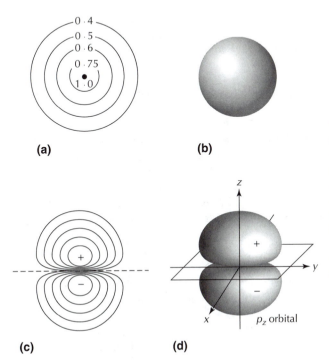

(a) **(b)** **(c)** **(d)** p_z orbital

Figure 3.8 (a) Contour map of ψ for a 1s orbital. (b) The spherical surface for a constant value of ψ for a 1s orbital. (c) Contour map of ψ in a plane through the z axis of the $2p_z$ orbital. (d) The two surfaces of constant ψ for the $2p_z$ orbital. Note the nodal surface in the xy plane. The upper surface corresponds to a positive value of ψ and the lower surface to a negative value of ψ. (Adapted with permission from R. McWeeny, *Coulson's Valence*, 1979, Oxford University Press, Oxford, Figures 2.3 and 2.7.)

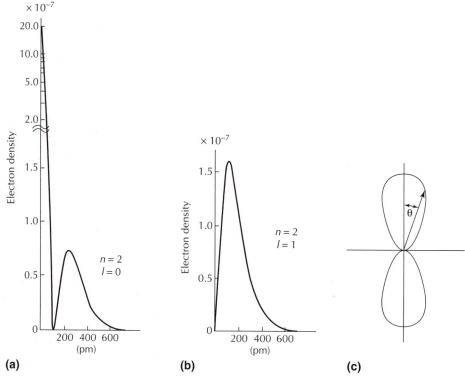

Figure 3.9 (a) Plots of the electron density R^2 along a radius for (a) a 2s orbital and (b) a 2p orbital. (c) Plots of $(\Theta\Phi)^2 = (\text{constant})\cos^2\theta$. (Reproduced with permission from J. E. Huheey, E. A. Keiter, and R. L. Keiter, Inorganic Chemistry, 4th ed., 1993, Harper Collins, Figures 2.2 and 2.6.)

of one of the outer contours as in Figure 3.10c. An alternative and useful method for depicting a three-dimensional electron density distribution is to use a two-dimensional dot density diagram, where the density of the dots is proportional to the electron density in the chosen plane, as in Figure 3.10a,d. The shape of ψ^2 is often called the orbital shape, although strictly speaking it is the shape of the electron density distribution for the orbital.

We have gone into the correct shapes of the functions ψ (orbitals) and ψ^2 (electron density distributions) in some detail because very approximate diagrams such as those in Figure 3.11 are often used to depict both ψ and ψ^2. For example, the diagrams in Figure 3.11 are used to depict both the shapes of 2p orbitals and their electron density distributions. By comparing Figures 3.7–3.11 we can see that the diagram in Figure 3.11 are only poor approximations to the true shapes of the functions ψ and ψ^2 for the 2p electrons. Moreover they do not show the rapid decrease in both ψ and ψ^2 in a radial direction. Although these depictions of orbital shapes are commonly used they can be quite misleading if their limitations are not understood as we will see, for example, in Section 3.10. We briefly discuss 3d orbitals, that are the orbitals for which $n = 3$ and $l = 2$ in Chapter 9.

Other atoms with only a single electron (He^+, Li^{2+}, etc.) are known as hydrogen-like atoms. The Schrödinger equation for such a system is the same as that for the hydrogen atom

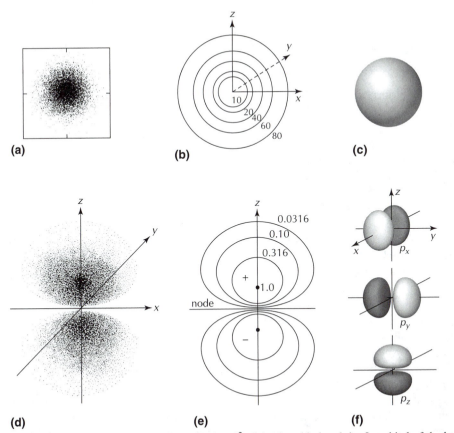

Figure 3.10 Representations of the electron density ψ^2 of the 1s orbital and the 2p orbital of the hydrogen atom. (b,e) Contour maps for the xz plane. (c,f) Surfaces of constant electron density. (a,d) Dot density diagrams: the density of dots is proportional to the electron density. (Reproduced with permission from the *Journal of Chemical Education 40*, 256, 1963; and M. J. Winter, *Chemical Bonding*, 1994, Oxford University Press, Fig. 1.10 and Fig. 1.11.)

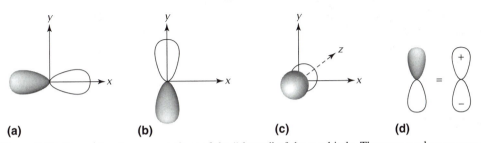

Figure 3.11 Conventional representations of the "shapes" of the p orbitals. They are used as approximate representations of both ψ and ψ^2. The sign of ψ is indicated by the shading. (Reproduced with permission from and M. J. Winter, *Chemical Bonding*, 1994, Oxford University Press, Oxford.)

except that the nuclear charge of $+1$ must be replaced by the appropriate atomic number Z. This decreases the size of the orbitals, but otherwise they have the same shape as the hydrogen atomic orbitals.

◆ 3.8   Electron Spin

The electron distribution in an atom or molecule containing more than one electron is determined by the electrostatic repulsion between the electrons and the attraction of the nuclei for the electrons. But there is another property of electrons that influences the electron density substantially, albeit in an indirect way. This property is called electron spin.

In 1925 Uhlenbeck and Goudsmit found that some puzzling features of atomic spectra could be explained only if it was assumed that an electron has an *intrinsic* angular momentum and therefore, because it is a charged particle, an intrinsic magnetic moment. This intrinsic angular momentum is a fundamental characteristic property of an electron, like its mass and its charge, but it is a quantum mechanical property that has no classical analogue. A useful model is to consider the electron as a sphere rotating about an axis through its center. Because a rotating charge produces a magnetic field, it therefore has a magnetic moment. This intrinsic magnetic moment is called **spin angular momentum** or simply **spin.** An electron can also have an angular momentum by virtue of its motion around the nucleus, which is called its orbital angular momentum.

Thus because of its spin, an electron behaves like a magnetic dipole. Classically, a magnetic dipole, such as the bar magnet in a compass, can have any orientation, from the lowest energy orientation in which it is pointing with the field to the highest energy orientation when it is pointing against the field (in this case the earth's magnetic field). But, as suggested by Uhlenbeck and Goudsmit, and confirmed experimentally by Stern and Gerlach (Box 3.2), spin angular momentum is quantized and can have just two values. We can imagine that the electron can rotate, or spin, either clockwise or counterclockwise about an axis in space, such as that provided by an external magnetic field, so that its magnetic moment lines up in the direction of the field or in the opposite direction. We should not, however, take this model literally because we have no way of observing an electron to see if it is a sphere or how it is spinning.

The magnetic forces between electrons are negligibly small compared to the electrostatic forces, and they are of no importance in determining the distribution of the electrons in a molecule and therefore in the formation of chemical bonds. The only forces that are important in determining the distribution of electrons in atoms and molecules, and therefore in determining their properties, are the electrostatic forces between electrons and nuclei. Nevertheless electron spin plays a very important role in chemical bonding through the Pauli principle, which we discuss next. It provides the fundamental reason why electrons in molecules appear to be found in pairs as Lewis realized but could not explain.

◆ 3.9   The Pauli Principle

The wave function for any system is a function of both the spatial coordinates of the electrons and of the spins of the electrons. It is convenient to describe the two possible values of the spin angular momentum of an electron as the two possible values of its spin coordinate,

▲ BOX 3.2 ▼
The Experimental Determination of Spin

The earliest experiment to determine the spin of an electron was carried out in 1921 by Stern and Gerlach. They passed a beam of neutral atoms, such as a Ag atom, which has a single unpaired electron with spin angular momentum but no orbital angular momentum, through an inhomogeneous magnetic field. Because classically we expect the magnetic dipoles to have a random orientation we would expect the beam of atoms to be spread out uniformly. Remarkably the beam *is split into just two beams* as we see in the figure. Because the two ends of the dipole are in slightly different field strengths, the dipole experiences a force either pulling it in the direction of the stronger field or repelling it away from this direction, depending on the orientation of the dipole. The magnetic moment calculated from the separation of the two beams is exactly equal to that postulated by Uhlenbeck and Goudsmit to explain various details of atomic spectra.

An even more direct proof would appear to be possible by passing a beam of electrons through such an inhomogeneous field. However, this experiment has never been successfully carried out because the effect is obscured by the large deflection produced by the action of the field on a moving charge. This complication is avoided when a beam of neutral atoms having one unpaired electron with no orbital magnetic moment is used.

The experimental arrangement for the Stern–Gerlach experiment is shown below. (a) A beam of atoms is passed through an inhomogeneous magnetic field produced by the specially shaped pole pieces. (b) The expected result for magnets that can take up any orientation with respect to the field: the beam is spread out uniformly. (c) The experimental result for silver atoms, which have only one unpaired electron in the valence shell. The beam is split up into two distinct beams. This result shows that the magnetic moment due to the single valence shell electron can take up only two orientations with respect to the field. (From R. J. Gillespie, D. A. Humphreys, N. C. Baird, and E. A. Robinson, *Chemistry,* 2nd ed., 1989, Allyn & Bacon, Boston.)

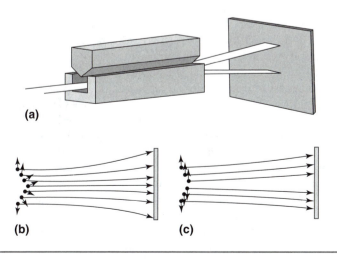

(a)

(b)   (c)

typically denoted as α and β. All electrons are identical, and therefore indistinguishable one from another. It follows that the interchange of the positions and of the spins of any two electrons in a system must leave the system unchanged. In particular, the probability distribution of the electrons, the electron density, must remain unchanged. In other words, ψ^2 must remain unchanged when the space and spin coordinates of any two electrons are interchanged. This requirement places a restriction on the wave function ψ itself. Either ψ must remain unchanged or it must change sign. We say that ψ must be either **symmetrical** or **antisymmetrical** to electron interchange. In fact, only antisymmetrical wave functions are found to represent the properties of electrons. That the wave function for any polyelectronic system must be anti-symmetric to electron interchange ($\psi \rightarrow -\psi$ on electron interchange) is a fundamental non-classical property of electrons. It shares this property with other particles such as protons, neutrons, and neutrinos, which have half-integral spins and are collectively called **fermions.** Particles, such as the α particle and the photon, that have integral spins (including zero) have symmetrical wave functions ($\psi \rightarrow \psi$ on particle interchange) and are called **bosons.**

The requirement that electrons have antisymmetrical wave functions is called the **Pauli principle,** which can be stated as follows:

> An electronic wave function must be antisymmetric with respect to the interchange of any two electrons.

No theoretical proof has been given of the Pauli principle. It is injected into current theories of electronic structure as a working rule. It represents some deep property of electrons, and indeed of all matter, which is not understood. A corollary of the Pauli principle is that no two electrons with the same spin can ever be at the same point in space. If two electrons of the same spin were at the same point in space, then on interchanging these two electrons, the wave function would remain unchanged and in particular would not change sign as is required by the Pauli principle. However, electrons with opposite spin can be at the same point in space, although the probability is very small owing to Coulomb repulsion. The main point, however, is that there are two worlds that do not see each other in terms of the Pauli principle: the set of α electrons and the set of β electrons. The only way these two sets can interact is via the Coulomb repulsion interaction.

The effect of the Pauli principle on the distribution of electrons is well illustrated by an example discussed by Lennard-Jones (1954). He considered the distribution of three particles with the same spin that are electron-like but have no charge, such that there are *no force fields* of any kind between them and that are confined to move in a circular ring. Because these uncharged electron-like particles do not repel each other, we can isolate the effect the Pauli principle alone has on this system. By considering three such particles located on a ring, Lennard-Jones showed that the effect of the Pauli principle is that in the most probable arrangement, the particles are arranged at equal intervals around the ring, that is, at the corners of an equilateral triangle. There is zero probability of any two of the particles being at the same point. It can also be shown that for any arrangement of two of the particles, the most probable position of the third is that in which it is as far as possible from the other two (Figure 3.12). In general we can say that electrons with the same spin keep as far apart as possible and have a zero probability of being found at the same point in space. In contrast, the Pauli principle has no effect on the relative arrangement of particles of opposite spin. Because electrons of the same spin keep as far apart as possible, they behave as if there were a repulsive force between them, which is sometimes called a Pauli force (Box 3.3).

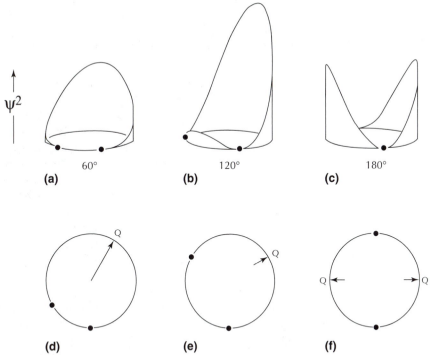

Figure 3.12 The distribution of a particle with zero charge constrained to move in a ring relative to two others of the same spin. Plots a–c show the probability ψ^2 of finding the third particle at a given location on the ring which is a maximum when all three particles are located at 120° intervals around the ring. In d–f Q indicates the most probable location of the third particle with respect to the other two. (Reproduced with permission from J. E. Lennard-Jones, *Adv. Sci.* 11, No. 54, 1956.)

The shell structure of an atom is also a consequence of the operation of the Pauli principle. As the charge on the nucleus increases with increasing atomic number, it would at first sight appear that the electrons would be pulled closer to the nucleus. This would lead to the atoms becoming smaller with increasing nuclear charge. But experiments show that they become *larger* as we proceed down any group of the periodic table. Is this increase in atomic size simply due to the increased number of electrons, which need more space as electrons repel each other? The answer is no, because it can be shown that the effect of an increasing nuclear charge more than compensates for the extra electron–electron repulsion resulting from any additional electrons, so we would expect atoms to shrink with increasing nuclear charge, which is not the case. No satisfactory explanation can be given on the basis of electrostatic forces only. Ultimately, it is the operation of the Pauli principle that prevents the electrons from all crowding toward the nucleus and leads to the formation of electron shells.

The Pauli principle is usually first met in a more restricted form called the Pauli exclusion principle, which states that

No more than two electrons may occupy one orbital, and if two electrons are present they must have opposite spins.

▲ BOX 3.3 ▼
The Pauli Force and Its Consequences

Electrons with the same spin behave as if there is a repulsive "force" acting between them. This apparent "force" is sometimes called the **Pauli force.** However, it is preferable not to speak of Pauli forces, since they are only apparent forces, not real forces like electromagnetic or gravitational forces. In fact, the Pauli principle implies that there is an intimate interconnection between the constituent parts of matter in the universe. Strictly speaking, no part can be isolated from the rest, except in an idealized way. The Pauli force acts at any time and over huge distances, much larger than atomic dimensions, but its effect becomes dramatic only when electrons of the same spin happen to be close to each other.

Pauli forces can be said to affect the energy of a molecule, but only indirectly. Kauzman has given a useful analogy: The Pauli principle determines the energy of a molecule in much the same way that traffic lights affect the number of automobile accidents. Strictly speaking, traffic lights do not prevent accidents, they merely influence the flow of traffic so that automobiles avoid one another. This decreases the number of collisions, hence reduces the number of accidents. For conciseness, however, we ordinarily say that traffic lights prevent accidents. Similarly, the movements of electrons are controlled in accordance with the Pauli principle, and these movements in turn affect the energy of a molecule so that there appears to be a force acting between electrons of opposite spin. The distribution of electrons in an atom or a molecule, hence the energy of the molecule and the electron distribution, are determined by the Pauli principle, together with electromagnetic forces.

Although the Pauli principle seems to be a very abstract concept, we do in fact have direct experience of it because it is responsible for the solidity of matter. According to our model of an atom in which a certain number of very small electrons are moving around a very tiny nucleus, it would appear that most of the space around the nucleus is empty. However, because of the Pauli principle, in any region of space

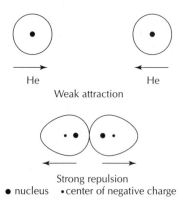

He   He
Weak attraction

Strong repulsion
● nucleus •center of negative charge

around a nucleus in which there is a high probability of finding a pair of electrons of opposite spin, there is only a low probability of finding any other electrons. Since most molecules have an equal number of electrons of opposite spin, no other electrons can penetrate significantly into the region occupied by these electrons, and thus molecules cannot penetrate into each other to a significant extent. Thus solid objects feel "hard" to our touch. Molecules that are brought closely together repel each other because their charge clouds cannot overlap to a significant extent.

The repulsion due to the Pauli principle can be illustrated by a very simple example: two helium atoms that are brought very close together as represented diagramatically in the figure. At relatively large distances apart there are only weak attractive forces between the two. However, when the atoms are brought close together the two charge clouds resist overlapping so that each charge cloud is distorted and pushed away from the other and they no longer surround each nucleus symmetrically. The center of action of the negative charge on each atom does not then coincide with the nuclear charge and it exerts a force on the nucleus pulling it away from the other nucleus. In other words, the two atoms repel each other.

The connection between the Pauli exclusion principle and the more general Pauli principle can be understood as follows. If two electrons with the same spin α were to occupy the same orbital ψ, the total wave function for the system would be written $\psi^\alpha(x_1, y_1, z_1)\, \psi^\alpha(x_2, y_2, z_2)$. Exchanging the coordinates of the two electrons changes the wave function to $\psi^\alpha(x_2, y_2, z_2)\, \psi^\alpha(x_1, y_1, z_1)$, which is the same as before. Since the wave function does not change sign, it is forbidden by the Pauli principle. Hence two electrons with the same spin cannot be described by the same wave function, or, in other words, an orbital cannot contain two electrons with the same spin. As we shall see, this form of the Pauli principle is used in describing atoms and molecules in terms of orbitals.

◆ 3.10 Multielectron Atoms and Electron Configurations

We use the solutions of the Schrödinger equation for a hydrogen-like atom—in other words, hydrogen-like orbitals—to give an approximate description of multielectron atoms. We add electrons two at a time to successive hydrogen-like orbitals of increasing energy. In accordance with the Pauli principle, only two electrons can be described by the same orbital, and they must be of opposite spin. This imaginary procedure is known as the **aufbau (building-up) principle.** It gives only an approximate description of an atom because the electrostatic repulsion between the electrons is ignored. Whereas in the hydrogen atom the energy of any given state depends only on the principal quantum number n, the energy of a state in a multielectron atom depends also on the quantum number l. This is because the electrons in inner orbitals shield the nucleus from the electrons in the outer orbitals. For example, a 2s electron penetrates the 1s shell to a greater extent than a 2p orbital, as can be seen from Figure 3.6. So a 2p electron is more shielded from the nucleus than a 2s electron and so has a higher

Table 3.1 Electron Configurations of the Elements

Period	Atomic Number	Element	Electron Configuration	Period	Atomic Number	Element	Electron Configuration
1	1	H	$1s^1$		52	Te	$[Kr]\,4d^{10}5s^25p^4$
	2	He	$1s^2$		53	I	$[Kr]\,4d^{10}5s^25p^5$
	3	Li	$[He]\,2s^1$		54	Xe	$[Kr]\,4d^{10}5s^25p^6$
	4	Be	$[He]\,2s^2$		55	Cs	$[Xe]\,6s^1$
	5	B	$[He]\,2s^22p^1$		56	Ba	$[Xe]\,6s^2$
2	6	C	$[He]\,2s^22p^2$		57	La	$[Xe]\,5d^1\quad 6s^2$
	7	N	$[He]\,2s^22p^3$		58	Ce	$[Xe]\,4f^2\quad 6s^2$
	8	O	$[He]\,2s^22p^4$		59	Pr	$[Xe]\,4f^2\quad 6s^2$
	9	F	$[He]\,2s^22p^5$		60	Nd	$[Xe]\,4f^4\quad 6s^2$
	10	Ne	$[He]\,2s^22p^6$		61	Pm	$[Xe]\,4f^5\quad 6s^2$
	11	Na	$[Ne]\,3s^1$		62	Sm	$[Xe]\,4f^6\quad 6s^2$
	12	Mg	$[Ne]\,3s^2$		63	Eu	$[Xe]\,4f^7\quad 6s^2$
	13	Al	$[Ne]\,3s^23p^1$		64	Gd	$[Xe]\,4f^75d^16s^2$
	14	Si	$[Ne]\,3s^23p^2$		65	Tb	$[Xe]\,4f^9\quad 6s^2$
3	15	P	$[Ne]\,3s^23p^3$		66	Dy	$[Xe]\,4f^{10}\quad 6s^2$
	16	S	$[Ne]\,3s^23p^4$		67	Ho	$[Xe]\,4f^{11}\quad 6s^2$
	17	Cl	$[Ne]\,3s^23p^5$	6	68	Er	$[Xe]\,4f^{12}\quad 6s^2$
	18	Ar	$[Ne]\,3s^23p^6$		69	Tm	$[Xe]\,4f^{13}\quad 6s^2$
	19	K	$[Ar]\,4s^1$		70	Yb	$[Xe]\,4f^{14}\quad 6s^2$
	20	Ca	$[Ar]\,4s^2$		71	Lu	$[Xe]\,4f^{14}5d^16s^2$
	21	Sc	$[Ar]\,3d^14s^2$		72	Hf	$[Xe]\,4f^{14}5d^26s^2$
	22	Ti	$[Ar]\,3d^24s^2$		73	Ta	$[Xe]\,4f^{14}5d^36s^2$
	23	V	$[Ar]\,3d^34s^2$		74	W	$[Xe]\,4f^{14}5d^46s^2$
	24	Cr	$[Ar]\,3d^54s^1$		75	Re	$[Xe]\,4f^{14}5d^56s^2$
	25	Mn	$[Ar]\,3d^54s^2$		76	Os	$[Xe]\,4f^{14}5d^66s^2$
	26	Fe	$[Ar]\,3d^64s^2$		77	Ir	$[Xe]\,4f^{14}5d^76s^2$
	27	Co	$[Ar]\,3d^74s^2$		78	Pt	$[Xe]\,4f^{14}5d^96s^1$
4	28	Ni	$[Ar]\,3d^84s^2$		79	Au	$[Xe]\,4f^{14}5d^{10}6s^1$
	29	Cu	$[Ar]\,3d^{10}4s^1$		80	Hg	$[Xe]\,4f^{14}5d^{10}6s^2$
	30	Zn	$[Ar]\,3d^{10}4s^2$		81	Tl	$[Xe]\,4f^{14}5d^{10}6s^26p^1$
	31	Ga	$[Ar]\,3d^{10}4s^24p^1$		82	Pb	$[Xe]\,4f^{14}5d^{10}6s^26p^2$
	32	Ge	$[Ar]\,3d^{10}4s^24p^2$		83	Bi	$[Xe]\,4f^{14}5d^{10}6s^26p^3$
	33	As	$[Ar]\,3d^{10}4s^24p^3$		84	Po	$[Xe]\,4f^{14}5d^{10}6s^26p^4$
	34	Se	$[Ar]\,3d^{10}4s^24p^4$		85	At	$[Xe]\,4f^{14}5d^{10}6s^26p^5$
	35	Br	$[Ar]\,3d^{10}4s^24p^5$		86	Rn	$[Xe]\,4f^{14}5d^{10}6s^26p^6$
	36	Kr	$[Ar]\,3d^{10}4s^24p^6$		87	Fr	$[Rn]\quad 7s^1$
	37	Rb	$[Kr]\,5s^1$		88	Ra	$[Rn]\quad 7s^2$
	38	Sr	$[Kr]\,5s^2$		89	Ac	$[Rn]\,6d^1\quad 7s^2$
	39	Y	$[Kr]\,4d^15s^2$		90	Th	$[Rn]\,6d^2\quad 7s^2$
	40	Zr	$[Kr]\,4d^25s^2$		91	Pa	$[Rn]\,5f^26d^17s^2$
	41	Nb	$[Kr]\,4d^45s^1$		92	U	$[Rn]\,5f^36d^17s^2$
	42	Mo	$[Kr]\,4d^55s^1$		93	Np	$[Rn]\,5f^46d^17s^2$
	43	Tc	$[Kr]\,4d^65s^1$	7	94	Pu	$[Rn]\,5f^6\quad 7s^2$
5	44	Ru	$[Kr]\,4d^75s^1$		95	Am	$[Rn]\,5f^7\quad 7s^2$
	45	Rh	$[Kr]\,4d^85s^1$		96	Cm	$[Rn]\,5f^76d^17s^2$
	46	Pd	$[Kr]\,4d^{10}$		97	Bk	$[Rn]\,5f^9\quad 7s^2$
	47	Ag	$[Kr]\,4d^{10}5s^1$		98	Cf	$[Rn]\,5f^{10}\quad 7s^2$
	48	Cd	$[Kr]\,4d^{10}5s^2$		99	Es	$[Rn]\,5f^{11}\quad 7s^2$
	49	In	$[Kr]\,4d^{10}5s^25p^1$		100	Fm	$[Rn]\,5f^{12}\quad 7s^2$
	50	Sn	$[Kr]\,4d^{10}5s^25p^2$		101	Md	$[Rn]\,5f^{13}\quad 7s^2$
	51	Sb	$[Kr]\,4d^{10}5s^25p^3$		102	No	$[Rn]\,5f^{14}\quad 7s^2$
					103	Lr	$[Rn]\,5f^{14}6d^17s^2$

energy. Similarly, all p orbitals have a higher energy than the corresponding s orbital and d orbitals have a still higher energy. The energies of the orbitals in a multielectron atom can be obtained experimentally from atomic spectra and ionization energies. The electron configurations of the atoms of the elements are given in Table 3.1.

◆ 3.11 Bonding Models

The nuclei in a molecule are held together by the electrons, so the electron density provides the fundamental basis for understanding bonding. Only relatively recently, however, has it become possible to obtain reasonably accurate electron density distributions for molecules of moderate size: either experimentally in a few cases by X-ray crystallography (Section 6.5) or, more commonly and conveniently, from a rather accurate computed wave function. Consequently we can now use the analysis and interpretation of electron distributions as a basis for understanding bonding, as we describe in Chapter 6. In the past it was usual to obtain an approximate wave function by some approximation method and then use this wave function as a basis for attempting to understand bonding. Two of these approximation methods—the **valence bond (VB) method** and the **molecular orbital (MO) method**—have dominated the discussion of the chemical bond. Today the molecular orbital method is the more popular. It is the method used for the vast majority of ab initio calculations (Section 3.12) because it is computationally simpler and more economical than the valence bond method. However, many chemists continue to use approximate valence bond descriptions of the bonding in a molecule.

The VB and the MO methods are rooted in very different philosophies of describing molecules. Although at the outset each method leads to different approximate wave functions, when successive improvements are made the two ultimately converge to the same wave function. In both the VB and MO methods, an approximate molecular wave function is obtained by combining appropriate hydrogen-like orbitals on each of the atoms in the molecule. This is called the **linear combination of atomic orbitals (LCAO)** approximation.

3.11.1 The Valence Bond Method

The VB method recognizes that characteristic lengths, energies, and force constants can be attributed to the bonds in very many molecules, as we discussed in Chapter 2. We begin with the atoms that form the molecule and pair up the unpaired electrons to form bonds. There may be several ways in which the electrons can be paired up. For example, the unpaired electrons in the $2p_x$ and $2p_y$ orbitals of the oxygen atom can each be paired with a single occupied hydrogen 1s orbital. We might also pair the two hydrogen 1s electrons to give the following pairing schemes:

$$
\begin{array}{cc}
\mathrm{O} & \mathrm{O} \\
2p_x \quad 2p_y & 2p_x \quad 2p_y \\
\diagup \qquad \diagdown & \\
1s \qquad\quad 1s & 1s\!-\!\!-\!\!-\!\!1s \\
\mathrm{H}^1 \qquad\quad \mathrm{H}^2 & \mathrm{H}^1 \qquad\quad \mathrm{H}^2
\end{array}
$$

Localized bonding orbitals are then constructed from a linear combination of the orbital on each of the paired atoms. To do this we use the **principle of maximum overlap,** which states

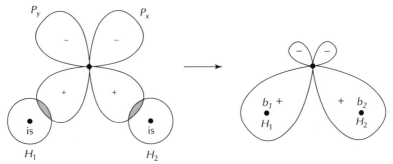

Figure 3.13 The formation of two bonding orbitals b_1 and b_2. (a) The overlap of the $2p_x$ and $2p_y$ orbitals of an oxygen atom with the 1s orbitals of two hydrogen atoms H_1 and H_2 leads (b) to the two bonding orbitals, b_1 and b_2.

that an AO on one atom should overlap as much as possible with an orbital on the atom to which it is bonded. The greater the overlap of the orbitals, the lower the energy of the molecule. Because the distance between the two H atoms is rather large, the overlap between the two 1s orbitals is small, so this structure does not make an important contribution and is usually neglected.

Figure 3.13 shows approximate representations of the $2p_x$ and $2p_y$ orbitals. To construct the two bonding orbitals b_1 and b_2 we place the H atoms along the directions of the $2p_x$ and $2p_y$ orbitals of the O atom to give maximum overlap of a 1s orbital of H with the $2p_x$ and $2p_y$ O orbitals (Figure 3.13). This bonding model implies a bond angle of 90° for the water molecule, which is not in very good agreement with the observed angle of 104.5°.

To enhance the overlap between an orbital on oxygen and the 1s orbital on hydrogen, and in general to obtain orbitals consistent with the geometry of the molecule, Pauling introduced the concept of **hybridization.** In this process, orbitals *centered on the same atom* are combined to obtain new hybrid orbitals, each of which is more concentrated in one direction than in the opposite direction and thus has a greater overlap with a ligand orbital (Figure 3.14).

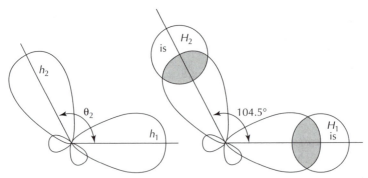

Figure 3.14 Hybrid orbitals h_1 and h_2 formed from the s and p orbitals of the oxygen atom to correspond to the bond angle of 104.5°. These orbitals have a greater overlap with the hydrogen 1s orbitals than the atomic 2p orbitals and so form stronger bonds.

For example, we can construct h_1 and h_2, two hybrid orbitals for describing the bonds in the water molecule, by taking the two combinations

$$h_1 = s + \lambda p_x \qquad h_2 = s + \lambda p_y \tag{3.16}$$

where λ is an adjustable mixing parameter. The value of λ determines the relative contributions of the s and p orbitals and also the angle between the hybrids, which is given by $1 + \lambda^2 \cos^2\theta$, where θ is the angle between the two hybrid orbitals. From the experimental value of 104.5° for the bond angle in the water molecule, we find $\lambda = 0.80$. Hence each of the two hybrid orbitals contain 80% p, character and 20% s character. This hybridization gives two orbitals that are more concentrated in the direction of the H atoms, and have a greater overlap with H_1, and H_2 than the p orbitals in Figure 3.13 and are consistent with the observed geometry.

Other hybrid orbitals that are commonly used in the valence bond description of bonds, particularly for organic molecules, are the sp, sp^2, and sp^3 hybrids. The sp orbitals, which have equal contributions from an s and a p orbital so that $\lambda = 0.5$ and the angle between the two hybrids is 180°, are used, for example, in the description of the bonds in the ethyne molecule. The sp^2 orbitals are constructed from equal contributions of an s and two p orbitals, and the sp^3 orbital from equal contributions of the s and three p orbitals (Table 3.2). They are used to describe the bonds in ethene and methane, where the bond angles are approximately 120° and exactly 109.5°, respectively. Figure 3.15 illustrates the formation of an sp hybrid orbital from the 2s and 2p orbitals and some popular representations of the shape of this orbital. We can see that the very commonly used depiction of an sp orbital in Figure 3.15d is only a very rough approximation. The electron density distributions ψ^2 of sp, sp^2, and sp^3 orbitals are shown in Figure 3.16 in the form of dot density diagrams. They all have rather similar shapes and are also quite similar to other shapes of diagrams of ψ (Figure 3.15). In particular they are more concentrated in one direction along their axis than in the opposite direction, and so they have a greater overlap with another orbital in this direction. Figure 3.17 shows the relative orientation of the sp, sp^2, and sp^3 sets of hybrid orbitals. The very approximate nature of these representations of hybrid orbitals and the corresponding electron density distributions becomes clear when we realize that the total electron density of four sp^3 hybrid orbitals is spherical, as is the total electron density of the 2s and three 2p orbitals. When the sp^3 orbitals are used to construct four bonding orbitals in the methane molecule by overlapping them with four hydrogen 1s orbitals, the electron density is more

Table 3.2 Examples of Hybrid Orbitals Used to Describe Equivalent Bonds

	Molecule			
	Ethyne	*Ethene*	*Methane*	*Water*
Hybrid set	2(sp)	3(sp^2)	4(sp^3)	2($s^{0.40}p^{1.60}$)
Hybrid	$s^{0.5}p^{0.5}$	$s^{0.33}p^{0.67}$	$s^{0.25}p^{0.75}$	$s^{0.20}p^{0.80}$
Bond angle	180°	120°[a]	109.5°	104.5°
λ[b]	1	$\sqrt{2}$	$\sqrt{3}$	2.0

[a]This description of the bonding orbitals is only an approximation: the experimental bond angle is 116°.

[b]Equation 3.16.

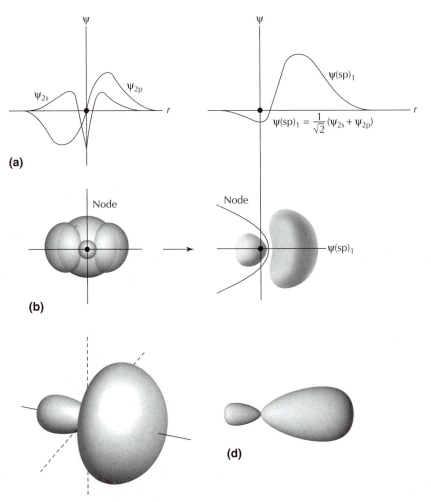

Figure 3.15 An sp hybrid orbital. (a) left, radial functions for the 2s and 2p atomic orbitals; right, radial function for the sp hybrid orbital (b) left, the shapes of the 2s and 2p atomic orbitals as indicated by a single contour value; right, the shape of the sp hybrid orbital as indicated by the same contour. (c) The shape of a surface of constant electron density for the sp hybrid orbital. (d) Simplified representation of (c). (Reproduced with permission from R. J. Gillespie, D. A. Humphreys, N. C. Baird, and E. A. Robinson, *Chemistry,* 2nd Ed., 1989, Allyn and Bacon, Boston.)

concentrated along each bond axis and the total density is no longer spherical. Thus, these approximate diagrams more closely represent the bonding orbitals in the molecules BeH_2, BH_3, and CH_4 than the hybrid orbitals from which they are constructed.

It is important to emphasize that:

• Hybridization is not a physical phenomenon. It is a mathematical operation that is used to construct localized orbitals to describe the bonding in a molecule.

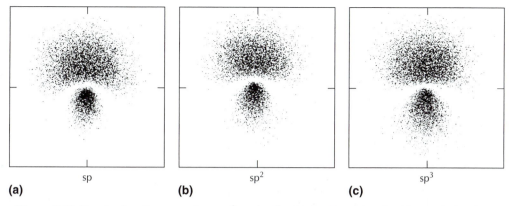

Figure 3.16 Dot density diagrams of sp, sp^2, and sp^3 orbitals. (Reproduced with permission from M. J. Winter, *Chemical Bonding,* 1994, Oxford University Press, Oxford.)

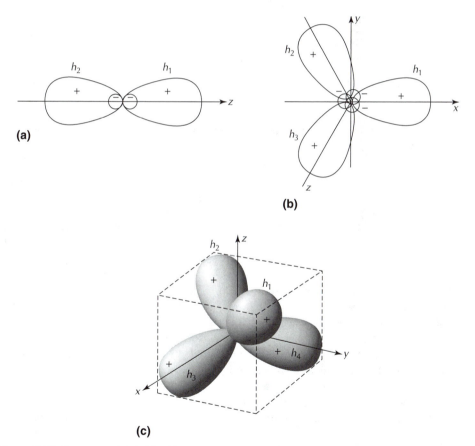

Figure 3.17 Geometry of hybrid orbitals. (a) digonal sp hybrids oppositely directed along the same axis; (b) trigonal sp^2 hybrids pointing along three axes in a plane inclined at 120°; (c) tetrahedral sp^3 hybrids pointing towards the corners of a regular tetrahedron. (Reproduced with permission from R. McWeeny, *Coulson's Valence,* 1979, Oxford University Press, Oxford.)

• Hybridization does not in general predict the equilibrium geometry. On the contrary, the construction of appropriate hybrid orbitals relies on a *given or predetermined* geometry.

For some molecules there are several reasonable pairing schemes. The total wave function is then a combination of the wave functions of each of these structures. For example, there are two equivalent pairing schemes for benzene corresponding to the two Kekulé structures. These pairing schemes contribute equally to the total wave function. These different possible valence bond structures correspond to Lewis structures and are the resonance structures mentioned in Chapter 1. The possibility of writing more than one pairing scheme implies that the electrons are not as localized as the pairing scheme or Lewis structure assumes. Such molecules are often more conveniently described by the molecular orbital method. Nevertheless, resonance structures are still often used in the description of bonding because of usefulness of the concept of localized bonds. However, when a large number of such structures are required for a satisfactory description, as is the case for the higher boranes, the method becomes cumbersome and less useful.

3.11.2 The Molecular Orbital (MO) Method

The central feature of the MO method is that orbitals are constructed from *all* the atomic orbitals in the valence shells of each atom that have the correct symmetry to have a net overlap. Unlike the orbitals of the valence bond method, which are restricted to two atoms only, the orbitals obtained by the MO method cover the whole molecule. These orbitals are then arranged in their presumed or calculated order of increasing energy and each is filled with two electrons of opposite spin in accordance with the Pauli exclusion principle and the aufbau principle. The use of atomic orbitals to construct molecular orbitals is, however, only a rough approximation. An accurate molecular orbital description can be obtained only by means of ab initio calculations (Section 3.12). The molecular orbital description of a molecule is particularly relevant for the description of electronic excitations in which the whole molecule participates, as in electronic spectra and photoionization. The energies of the molecular orbitals of a molecule can be obtained from such spectra. It is not possible to say that an excited electron came from a particular localized orbital because the whole molecule is affected by the ionization or removal of an electron. It is sometimes claimed, because such spectra can be described only in terms of an MO description, that the MO description of a molecule is superior to the VB description, but this is not the case. We must use whichever description is relevant to the phenomenon under discussion. If we are talking about bond properties such as length and energy and about geometry, then a localized orbital (VB) description is more relevant than an MO description. Consequently we do not make much use of the molecular orbital method in this book.

Double and triple bonds, particularly those in carbon molecules, are often described in MO terms. Thus a double bond is described as consisting of a σ bond formed by the overlap of an sp hybrid orbital on each carbon atom and a π bond formed by the "sideways" overlap of either the $2p_x$ or $2p_y$ orbitals (Figure 3.18). A σ orbital has cylindrical symmetry like an atomic s orbital whereas a π orbital, like an atomic p orbital, has a planar node passing through the nucleus of each of the bonded atoms. A triple bond is similarly described as consisting of a σ orbital and two π orbitals formed from both the $2p_x$ and $2p_y$ orbitals on

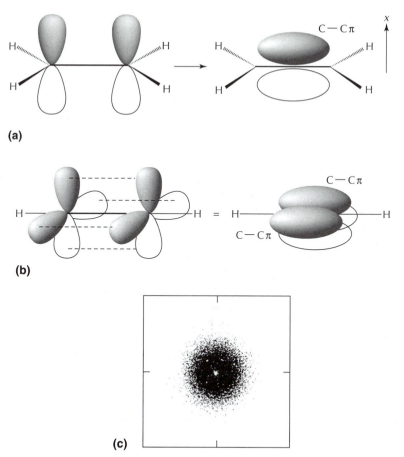

Figure 3.18 Conventional representations of π orbitals. (a) The π orbital in the ethene molecule. (b) The two π orbitals in the ethyne molecule. (c) A dot density diagram of the electron density in a plane perpendicular to the bond axis in ethyne. (Reproduced with permission from M. J. Winter, *Chemical Bonding,* 1994, Oxford University Press, Oxford.)

each carbon atom. (Figure 3.18). The inadequacy of the conventional representations of p orbitals is clear from these diagrams—they do not show any "sideways" overlap of the p orbitals, which is sometimes artificially represented by dashed lines as in Figure 3.18b. Nor do they show that the total π-electron density around the CC bond axis in ethyne is cylindrically symmetrical. These π orbitals are strictly speaking not molecular orbitals because they are localized to just two carbon atoms. However, the terminology of the MO method is used to distinguish this description from the alternative description that can be given in terms of localized orbitals.

This alternative description follows from classical ideas and from a VB description utilizing hybrid orbitals. According to this description, a double bond is described as consisting of two bent bonds, sometimes called τ bonds or banana bonds, formed by the overlap of

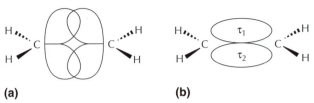

(a) (b)

Figure 3.19 Bent-bond representation of the double bond in ethene. The overlap of sp³ orbitals on each carbon atom produces to bend bond (τ) orbitals.

sp³ hybrid orbitals (Figure 3.19) and a triple bond as consisting of three bent or τ bonds. The very approximate conventional illustrations of bent bonds do not give a good picture of the total electron density, which, in ethene, as we will see in Chapter 6, has a maximum value along the bond axis and an elliptical cross section. The MO and VB descriptions are equivalent because the set of σ and π orbitals can be transformed into τ orbitals by forming appropriate linear combinations, as shown in Figure 3.20. We may use whichever description is the most convenient for our purposes. The σ–π description is the most popular because it can be easily extended to other unsaturated hydrocarbons and their derivatives (e.g., butadiene, benzene), as discussed in any organic chemistry textbook. Note, however, that it is strictly incorrect to say that a C=C double bond *consists* of a σ and a π orbital, or that it consists of two τ bonds—these are only convenient *descriptions*. It is also strictly incorrect to define a bond as being formed by the overlap of atomic or hybrid orbitals. This is simply a convenient *description* of a hypothetical process.

The MO and VB methods provide alternative but equivalent descriptions of the bonding in a molecule. A set of molecular orbitals can always be transformed into a corresponding set of more localized orbitals, and vice versa. For example, according to the MO de-

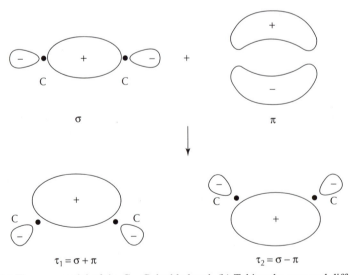

Figure 3.20 (a) The σ–π model of the C=C double bond. (b) Taking the sum and difference of these orbitals produces two bent bond (τ) orbitals.

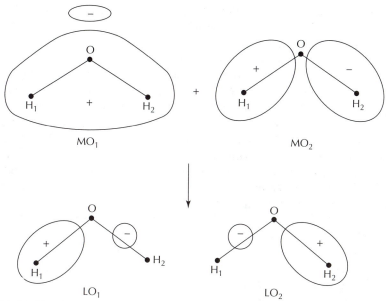

Figure 3.21 (a) Simplified representations of the two main bonding MOs of the water molecule. (b) Simplified representations of the localized equivalent bonding MOs in water.

scription, the water molecule has two bonding molecular orbitals whose approximate shapes are shown in Figure 3.21a. By taking two linear combinations of these two orbitals, that is, the sum and difference, we form two orbitals that are largely localized in each of the OH bonding regions and correspond approximately to the two localized bonding orbitals of the VB method (Figure 3.21b). We can think of fully localized orbitals and molecular orbitals as two limiting models. In real molecules electrons are not as localized as localized bonding models assume, nor are they as fully delocalized as the simple MO theory indicates.

◆ 3.12 Ab Initio Calculations

Thus far we have discussed only the most rudimentary forms of the valence bond and molecular orbital descriptions of molecules, which are nevertheless widely used by many chemists in the qualitative discussion of bonding and geometry. However, both these methods have been highly developed to obtain very accurate solutions of the Schrödinger equation and thus accurate, geometries energies and wave functions for even rather large molecules. Commercial quantum chemistry packages for carrying out these calculations have been available for some time. Their user-friendliness and high level of automation have encouraged many researchers to run them on personal computers and the results of such calculations are now a valuable complement to experiment. They are becoming increasingly regarded as "black boxes," just as experimental techniques are sometimes regarded. So it is not essential to have a detailed understanding of the theory but it is important to have, at least, some understanding of the reliability, accuracy, and limitation of the calculations. Because, in contrast to what are known as **semiempirical** calculations, these calculations do not make use of any experimental parameters but only the fundamental properties of electrons and nuclei, they are called **ab initio** (from first principles) calculations. We do not discuss semiempirical calculations, which are more easily carried out, but less accurate, because, although important in the past, they are now less important than ab initio calculations.

In the following we give only a brief and nonmathematical description of the development of the MO method to obtain geometries and wave functions. Since it is computationally both more simple and more economical to carry out ab initio calculations within the MO formalism (than working with the VB formalism), the MO method is the most commonly used.

Molecules in their ground state are typically treated using the so-called **Born–Oppenheimer** approximation. This approximation is also known as the **clamped nuclei** approximation because it views the electrons as moving in a field of fixed nuclei. In other words, the total wave function, which is a function of nuclear and electronic coordinates, can be separated into a nuclear wave function and an electronic wave function. This approximation can be justified on the basis that electrons move much faster than nuclei and follow them quasi-instantly.

The next important approximation in obtaining the wave function for an N-electron molecule is to write the N-electron wave function as a product of N one-electron wave functions or orbitals. This product is called the **Hartree** product. In this approximation the electrons do not see each other as individuals but only in an average way. More precisely, the probability of finding one electron in a small volume element is independent of the position of another individual electron. However, the Hartree product is not a realistic wave function because it does not obey the Pauli principle. The Hartree product must be **antisymmetrized** so that it changes sign when the labels of any two electrons are interchanged. The simplest way to do this is to write the wave function in the form of a single determinant called a **Slater determinant,** because a determinant has the convenient property of changing sign when any two of its columns are interchanged. A popular computational scheme for obtaining an approximate wave function in this way is called the **Hartree-Fock method.** Because this method replaces the instantaneous location of all the electrons other than the one considered by an average electron distribution, it neglects the **correlation** of the motion of the electrons due to their electrostatic repulsion. Correcting for this discrepancy remains a major challenge of quantum chemistry, as we discuss later.

The solution of the Hartree–Fock wave equation yields a set of Hartree–Fock orbitals, each with a corresponding orbital energy. For a system with N electrons in N orbitals, the N/2 lowest energy orbitals are called **occupied** orbitals, because each contains two electrons. The remaining members of the set are called **virtual** or **unoccupied** orbitals. The Hartree–Fock equation contains the potential due to the average distribution of electrons, which itself depends on the solution of the Hartree–Fock equation. To break this vicious circle, we start with approximate orbitals that are based on reasonable assumptions, from which we obtain an approximate potential enabling us to solve the Hartree–Fock equation. This procedure yields improved orbitals, which are different from the approximate orbitals we first used to construct the potential. We then use these improved orbitals to obtain a new potential from which, by solving the Hartree–Fock equation, we again obtain new orbitals, and so on. We repeat the procedure until the orbitals obtained differ insignificantly from those that went into the Hartree–Fock equation in the preceeding step. In other words, we continue the procedure until it ultimately converges iteratively to a self-consistent set of orbitals, which is why this method is called the **self-consistent field** method.

In practice, the Hartree–Fock equation is solved by starting with a set of **basis functions.** Originally basis functions were simply hydrogen-like orbitals, but modern calculations use linear combinations of highly modified orbitals as basis functions. The design of adequate basis functions is more driven by computational efficiency than by physical arguments. The larger and more complete the set of basis functions, the lower the Hartree–Fock energy becomes until a limit is reached, called the **Hartree-Fock limit.**

The remaining difference between the exact energy and the energy corresponding to the Hartree–Fock energy is called the **correlation energy,** and it is due to the electrostatic interaction between the electrons. Various methods have been devised to calculate this energy and thus approach more closely to the exact energy. All these methods, which are known by various acromyms (e.g., MP2, CCSD, QCISD, IPA), lead to some improvement in the accurate description of molecules, but at varying computational cost. Details can be found in several standard texts.

Some alternative computational methods such as **density functional theory (DFT)** bypass the Hartree–Fock approximation. The basic idea behind DFT is that the energy of an electronic system can be expressed solely in terms of the electron density (i.e., without reference to orbitals). The energy is then a "function" of the function ρ. Such a "function" is correctly called a *functional*. However, its exact form is not known. Various methods have been proposed for constructing an approximate functional from which an improved electron density and energy can be obtained and the calculation repeated iteratively until the calculated electron density and the corresponding energy cease to change significantly, as in the Hartree–Fock method. Currently DFT is a very active area of research and is being increasingly used in quantum chemistry. Many improved functionals have been designed that give rather accurate wave functions and energies, as judged by the agreement between the minimum energy geometry and the experimental geometry.

One of the major goals of modern quantum chemistry has been the calculation of molecular energies to an accuracy of ± 5–10 kJ mol^{-1}, which is the accuracy to which molecular energies can be obtained experimentally, and this goal has been attained in recent years. One of the techniques to obtain this accuracy is called **Gaussian 2 (G2).** G2 is a technique in which the energy of the molecule is incrementally decreased by increasing the number of basis functions, by using successively more refined methods to compute the correlation energy, and by the adjustment of the geometry of the molecule. In this way the energy of the lowest energy equilibrium geometry of the molecule is obtained, together with the corresponding wave function.

Determination of the geometry in this way is now a valid alternative to experimental methods. It is particularly useful for molecules that are so reactive or unstable that their geometry cannot be easily determined and for molecules that have not yet been observed.

◆ 3.13 Postscript

Ab initio calculations give us a knowledge of the energy of a molecule in its equilibrium ground state and its corresponding geometry and wave function. Unless it is a very reactive or unstable molecule, its geometry can also be determined experimentally and provides a good check on the accuracy of the calculated geometry. The wave function, however, which is not a physical observable, cannot be determined by experiment. Moreover, it does not directly give us any understanding of the bonding or geometry of a molecule. To attempt to obtain such an understanding we need to interpret the wave function. The total wave function is composed of N two-electron molecular orbitals each of which extends over the whole molecule. In order to interpret the wave function, it is usual to convert the molecular orbitals into equivalent, more localized, orbitals that we can hope to identify with concepts such as bonds and lone pairs. There are several criteria on the basis of which orbitals can be localized and the choice of procedure is arbitrary so the localized orbitals thus obtained are not unique and they are also not physically observable.

Alternatively, we can base our analysis on the electron density, which as we have seen, is readily obtained from the wave function. The advantage of analyzing the electron density is that, unlike the wave function, the electron density is a real observable property of a molecule that, as we will see in Chapter 6, can be obtained from X-ray crystallographic studies. At the present time however, it is usually simpler to obtain the electron density of a molecule from an ab initio calculations rather than determine it experimentally. Because this analysis is based on a real physically observable property of a molecule, this approach appears to be the more fundamental. It is the approach taken by the atoms in molecules (AIM) theory, which we discuss in Chapters 6 and 7, on which we base part of the discussion in Chapters 8 and 9.

Before discussing the AIM theory, we describe in Chapters 4 and 5 two simple models, the valence shell electron pair (VSEPR) model and the ligand close-packing (LCP) model of molecular geometry. These models are based on a simple qualitative picture of the electron distribution in a molecule, particularly as it influenced by the Pauli principle.

▶ Further Reading

Elementary Introductions to Quantum Chemistry and the Chemical Bond

P. A. Cox, *Introduction to Quantum Theory and Atomic Structure,* 1996, Oxford University Press, Oxford.

H. B. Gray, *Chemical Bonds: An Introduction to Atomic and Molecular Structure,* 1994, University Science Books, Mill Valley, CA, 1994.

M. J. Winter, *Chemical Bonding,* 1994, Oxford University Press, Oxford.

More Advanced Books on Quantum Chemistry and Chemical Bonding

P. W. Atkins, *Quanta. A Handbook of Concepts,* 2nd Ed, 1991, Oxford University Press, Oxford.

I. N. Levine, *Quantum Chemistry,* 5th ed., 2000, Allyn & Bacon, Boston.

J. P. Lowe, *Quantum Chemistry,* 2nd ed., 1993, Academic Press, San Diego.

R. McWeeny, *Coulson's Valence,* 3rd ed., 1979, Oxford University Press, Oxford.

F. L. Pilar, *Elementary Quantum Chemistry,* 2nd ed., 1990, McGraw Hill, New York.

B. Webster, *Chemical Bonding Theory,* 1990, Blackwell, Edinburgh.

Theoretical and Computational Chemistry

J. B. Foresman and A. Frisch, *Exploring Chemistry with Electronic Structure Methods,* 2nd ed., 1996, Gaussian, Inc. Pittsburgh.

A practical guide and tutorial to the ab initio program Gaussian.

W. J. Hehre, L. Radom, P.v.R. Schleyer and J. A. Pople, *Ab intio Molecular Orbital Calculations,* 1986, Wiley, New York.
An older but still valuable introduction to ab initio calculations.

F. Jensen, *Introduction to Computational Chemistry,* 1999, Wiley, Chichester.
A medium level text covering a wide variety of topics.

A. Hinchcliffe, *Modeling Molecular Structures,* 1996, Wiley, Chichester.
An easily digestable account of practical molecular modeling with any examples and computer outputs.

E. S. Kryachko and E. V. Ludena, *Energy Density Functional Theory of Many-Electron Systems,* 1990, Kluwer, Dordrecht.
Extensive and detailed standard work on all aspects of DFT.

R. McWeeney, *Methods of Molecular Quantum Mechanics,* 2nd ed., 1992, Academic Press, San Diego.
Advanced mathematically oriented classic work.

R. Parr and W. Yang, *Density Functional Theory,* 1989, Oxford University Press, New York.
Classic text on DFT.

A. Szabo and N. S. Ostlund, *Modern Quantum Chemistry, Introduction to Advanced Electronic Structure Theory,* 1st ed. revised, 1989, McGraw Hill, New York.
Well-written classic text on the core aspect of modern ab initio calculations containing many useful exercises.

▶ References

Gaussian 94 An ab initio program written by M. J. Frisch, G. W. Trucks, H. B. Schlegel, P. M. W. Gill, B. G. Johnson, M. A. Robb, J. R. Cheeseman, T. Keith, G. A. Petersson, J. A. Montgomery, K. Raghavachari, M. A. Al-Laham, V. G. Zakrzewski, J. V. Ortiz, J. B. Foresman, J. Cioslowski, B. B. Stefanov, A. Nanayakkara, M. Challacombe, C. Y. Peng, P. Y. Ayala, W. Chen, M. W. Wong, J. L. Andres, E. S. Replogle, R. Gomperts, R. L. Martin, D. J. Fox, J. S. Binkley, D. J. Defrees, J. Baker, J. P. Stewart, M. Head-Gordon, C. Gonzalez, and J. A. Pople, 1995 Gaussian, Inc., Pittsburgh.

J. E. Lennard-Jones, *Adv. Sci. 11 (No. 54),* 136, 1954.

W. Kauzmann, *Quantum Chemistry: An Introduction,* 1957, Academic Press, New York.

4

C H A P T E R

MOLECULAR GEOMETRY AND
THE VSEPR MODEL

■ ■ ■

◆ 4.1   Introduction

Stereochemistry, the study of the three-dimensional structures of molecules, was born in 1874 with the independent proposals by le Bel and van't Hoff that the four valences of carbon are directed toward the corners of a tetrahedron. By means of this model they were able to account for the first time for the existence of chiral molecules. Their model was later extended to other four-coordinated atoms such as silicon, to boron in ions such as BF_4^-, and to nitrogen in ions such as NH_4^+. In his coordination theory Werner proposed in 1893 that six-coordinated transition metal molecules and ions have an octahedral geometry. He proved this in 1911 by separating the optical isomers of $[Co(en)_3]Cl_3$ and similar compounds. The tetrahedral geometry of many four-coordinated atoms and the octahedral geometry of many six-coordinated atoms was clearly established by the early 1900s. Little information about other molecular geometries, and no quantitative information, was available until the 1930s. The development of physical methods for the direct determination of molecular geometry such as X-ray and electron diffraction, and infrared and Raman spectroscopy in the 1930s led to the confirmation of the tetrahedral geometry of most four coordinated molecules and the octahedral geometry of most six-coordinated molecules. The trigonal bipyramidal geometry of some five-coordinated molecules was also established, and a few examples of square planar four-coordinated molecules were discovered. More important, however, it became possible for the first time to measure **bond lengths,** the distance between the nuclei of two atoms that are bonded together, **bond angles,** the angles between bonds to the same atom, and **torsional angles,** the angles between bonds on adjacent atoms. These are the only properties of bonds that can be clearly defined and measured for bonds of all types in all molecules. These properties are therefore particularly useful for discussing the nature of bonds. In this chapter we discuss molecular shapes and bond angles. Bond lengths and other related bond properties were discussed in Chapter 2.

In 1940 Sidgwick and Powell surveyed the geometry of the singly bonded AX_n molecules whose structures were known at the time and showed that most of them could be rationalized on the assumption that the geometry of an AX_n molecule is determined by the total number of electron pairs, both bonding and nonbonding, in the valence shell of the central atom A, as given by the Lewis structure for the molecule. They concluded that two pairs of electrons in a valence shell have a linear arrangement, three pairs a triangular arrangement, four pairs a tetrahedral arrangement, five pairs a trigonal bipyramidal arrangement, and six pairs an octahedral arrangement. The paucity of structural information on molecules with higher coordination numbers prevented them from reaching any firm conclusions about the preferred arrangements of more than six electron pairs in a valence shell. They included transition metal molecules in their discussion but noted that among these molecules there were many exceptions to their generalizations.

Sidgwick and Powell's important proposal did not gain much attention until it was developed and extended by Gillespie and Nyholm in 1957 into what has subsequently become known as the **VSEPR (valence shell electron pair repulsion) model** or sometimes as the Gillespie–Nyholm rules. Gillespie and Nyholm were able to discuss a considerably larger number of molecules, and they showed that the geometry of the vast majority of the molecules of the main group elements, particularly those of the nonmetals, is consistent with the electron pair arrangements proposed by Sidgwick and Powell. They showed that these electron pair arrangements are those that maximize the distances between the electron pairs and that they are primarily a consequence of the operation of the Pauli principle. Moreover, by considering the differences between lone pairs and bond pairs, and between single, double, and triple bonds, as well as the effects of ligand electronegativity, they were able to give a qualitative explanation of deviations of bond angles from the ideal values of 90, 109.5, and 120° found in molecules with structures based on regular polyhedra. Since that time the number of molecules whose geometry can be rationalized by the VSEPR model has grown enormously. Moreover, the model itself has undergone some modifications and improvements, and its limitations have been more clearly understood.

This chapter is devoted to an account of the VSEPR model particularly as applied to molecules of the main group elements.

◆ 4.2   The Distribution of Electrons in a Valence Shell

The fundamental assumption of the VSEPR model is that electron pairs adopt arrangements that keep them as far apart as possible. They behave *as if* they repel each other. As we saw in Chapter 3, this behavior is mainly a consequence of the Pauli principle and is not primarily a result of electrostatic repulsion. We discussed in Chapter 3 the effect of the Pauli principle on the distribution of same-spin electrons. We saw that if three electrons of the same spin between which there are no force fields of any kind are constrained to move in a circular ring their most probable arrangement is the one in which they are at equal intervals around the ring—in other words, as far apart as possible. This arrangement is reinforced by the electrostatic repulsion between the electrons, but electrostatic repulsion it is not the main reason for the adoption of this most probable arrangement, since this is determined primarily by the Pauli principle. The argument can be extended to four electrons confined to move on the surface of a sphere. Their most probable arrangement, which is due primarily to the operation of the Pauli principle reinforced by electrostatic repulsion, is that in which they are at the vertices of a tetrahedron. Now we are ready to consider why electrons in molecules appear to be found in pairs that are also arranged as far apart as possible.

The formation of electron pairs in molecules cannot be understood in terms of electrostatics alone, as Lewis realized but could not explain; it can be understood, however, if both electrostatics and the Pauli principle are taken into account. In most molecules there are equal numbers of electrons of opposite spin. We first consider the very common case of a valence shell containing eight electrons, four of α spin and four of β spin, and we assume that they are at the same average distance from the nucleus. In accordance with the Pauli principle, in the most probable *relative* arrangement of the four α electrons they are as far apart as possible—in other words, they are at the vertices of a tetrahedron (Figure 4.1). In a free atom or ion, such as Ne, F^-, or O^{2-}, this most probable *relative* tetrahedral arrangement of the α electrons does not have a fixed orientation. Thus there is an equal probability of finding an α electron anywhere in the valence shell; in other words, the total electron density distribution of the four α electrons is spherical. Similarly, the most probable arrangement of the four β electrons is at the vertices of a tetrahedron, which in a free atom or ion does not have any fixed orientation in space. Thus their total electron density distribution is also spherical. From

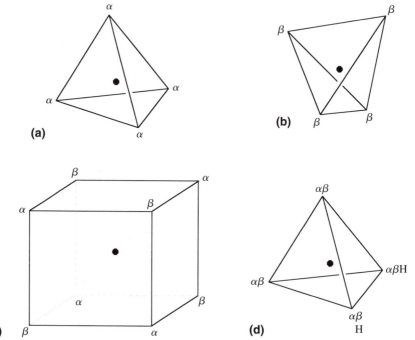

Figure 4.1 The most probable relative arrangement of (a) four α-spin electrons, (b) four β-spin electrons, (c) four α- and four β-spin electrons, and (d) four pairs of $\alpha\beta$-spin electrons in the H_2O molecule.

the point of view of the Pauli principle, there is no preferred arrangement of one spin set relative to the other. However, electrostatic repulsion, while reinforcing the tetrahedral arrangement of each spin set, keeps the tetrahedra apart, so that the most probable *relative* arrangement of the two spin sets is that in which they are occupying alternate corners of a cube (Figure 4.1c). Nevertheless, because this relative arrangement of the electrons does not have any fixed orientation in space, the overall total charge distribution is spherical. It is interesting to note that Lewis first proposed that the eight electrons of an octet have a cubic arrangement, but he later abandoned this model for a tetrahedral arrangement of four pairs. Lewis's first suggestion was essentially correct for a free atom or ion. As we shall see, *electron pairs are present only in molecules, not in free atoms or ions.*

In a molecule the spherical electron distribution of a free atom is perturbed by the atoms with which it is combined. Let us imagine, for example, that we start with a spherical oxide ion O^{2-} and form a water molecule by bringing up two protons. Each proton attracts electrons, but, as a consequence of the Pauli principle, only two electrons of opposite spin can be attracted toward each hydrogen nucleus. *If the hydrogen nuclei attract the oxygen valence shell electrons sufficiently strongly,* the tetrahedron of α electrons and the tetrahedron of β electrons will be brought into approximate coincidence, forming four electron pairs. Thus a pair of electrons has a high probability of being found in each of the regions between the oxygen core and a hydrogen nucleus. These two pairs are the bonding pairs. At the same time, two nonbonding, or lone, pairs of electrons are also necessarily formed, so that there are four electron pairs with an approximately tetrahedral arrangement—two bonding pairs and two nonbonding or lone pairs (Figure 4.1d). Now the four regions in which there is a high probability of finding a pair of electrons are fixed in space relative to the positions of the H nuclei, so the overall electron density around the oxygen is no longer spherical but is greater in the four tetrahedral directions than in other directions.

It is important to appreciate that two electrons of opposite spin do not attract each other to form a pair. Rather, despite their mutual repulsion, they are brought into the same region of space by the attraction of the positively charged ligand atom core. It is only two electrons of opposite spin that can be brought together in this way. A third electron has a very low probability of being found in this region, because it will have the same spin as one of the electrons of the bonding pair. We see, therefore, why a bond is always associated with a pair of electrons, as Lewis first realized but was unable to explain.

Similarly, if we bring up three protons to an N^{3-} ion, we form the triangular pyramidal NH_3 molecule with a tetrahedral arrangement of three bonding pairs and one nonbonding pair, and if we bring up four protons to a C^{4-} ion we form the tetrahedral CH_4 molecule (Figure 4.2). However, if a proton combines with an F^- ion, two electrons of opposite spin are brought together to form a bonding pair, but the two tetrahedra of same-spin electrons are free to rotate around the HF bond axis so that nonbonding pairs are not formed. Instead, the six nonbonding electrons are equally distributed in a ring perpendicular to the bond axis (Figure 4.3). In general, in linear molecules, electrons of opposite spin are not all brought together in pairs. Only those opposite-spin electrons that have a most probable location on the molecular axis are brought into coincidence, leaving the remaining electrons equally distributed in a ring around the molecular axis. That some of the electrons are not paired in linear molecules provided the basis for the double-quartet model proposed by Linnett in 1964. We discuss this model in Section 4.5.

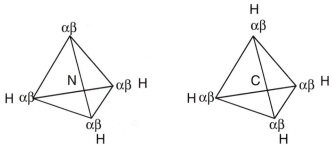

Figure 4.2 The tetrahedral arrangement of four $\alpha\beta$ pairs of electrons in the NH_3 and CH_4 molecules.

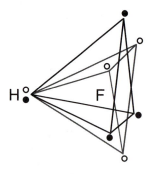

Figure 4.3 The most probable arrangement of the electrons in the valence shell of fluorine in the HF molecule. Only one localized pair is formed with the other electrons equally distributed in a ring behind the F nucleus. ● α spin electron or ○ β spin electron.

We see that it is a consequence of the Pauli principle and bond formation that the electrons in most molecules are found as pairs of opposite spin—both bonding pairs and nonbonding pairs. The Pauli principle therefore provides the quantum mechanical basis for Lewis's rule of two. It also provides an explanation for why the four pairs of electrons of an octet have a tetrahedral arrangement, as was first proposed by Lewis, and why therefore the water molecule has an angular geometry and the ammonia molecule a triangular pyramidal geometry. The Pauli principle therefore provides the physical basis for the VSEPR model.

◆ 4.3 Electron Pair Domains

So far we have considered just the most probable angular arrangements of electrons in a valence shell. But each electron can be found at locations other than its most probable location with decreasing probability with increasing distance from its most probable location. Each electron can be described by a charge cloud which is most dense at the most probable location of the electron and becomes less dense with increasing distance from this most probable location. Since both electrons of a bonding or nonbonding electron pair have the same probability distribution, they may be considered to form one two-electron charge cloud. In accordance with the Pauli principle, there is a low probability of finding any other electrons in the space that this electron pair occupies, and the more strongly this electron pair is localized, the more strongly it excludes other electron pairs from its space. When electron pairs are sufficiently localized, it is a useful, although rough, approximation to assume that each charge cloud almost completely excludes other charge clouds. We call the region of space occupied by a pair of electrons, that is, by the charge cloud of an $\alpha\beta$ pair of electrons, an **electron pair domain**. An electron pair domain can be defined as the region of space in which there is a high probability of finding an electron pair, or in which a large fraction of an electron pair charge cloud is found. An electron pair domain surrounds the point at which there is a maximum probability of finding a pair of electrons, that is, at which the electron charge cloud is most concentrated. The density of the charge cloud is a maximum at this point and the density decreases on moving away from this point. So in the water molecule there are four electron pair domains that have a tetrahedral arrangement, two bonding domains, and two lone pair domains.

We can see now why the static model of Lewis with four electron pairs in a tetrahedral arrangement is so useful even though electrons are in rapid motion and do not occupy fixed positions. Each electron pair occupies a reasonably well localized domain, and four domains have a tetrahedral arrangement.

We have so far considered valence shells containing four pairs of electrons, but we can extend the same arguments to other numbers of valence shell electron pairs. The most probable arrangements of pairs of opposite spin electrons in the valence shell of an atom in a molecule are two pairs, collinear; three pairs, equilateral triangular; four pairs, tetrahedral; five pairs, trigonal bipyramidal; six pairs, octahedral. This is because, as we will now see, these are the arrangements that keep the electron pairs as far apart as possible. We discuss valence shells with more than six electron pairs in Chapter 9.

These electron pair domain arrangements can be modeled in several simple ways. For example, they are the arrangements of a given number of points on the surface of a sphere in which each point is at a maximum distance from its neighbors, and each point represents a pair of electrons. This is the **points-on-a-sphere model** (Figure 4.4). The arrangements of two, three, four, and six points, namely, linear, triangular, tetrahedral, and octahedral, are intuitively obvious and easily proved. But the argument for the trigonal bipyramidal arrangement of five points is not quite as simple, and its treatment is deferred until Section 4.6 The same arrangements are obtained by the **circles-on-a-sphere model** in which a given number of equal circles are packed on the surface of a sphere so that they occupy a maximum area of the surface (Figure 4.5). Yet another useful model was first suggested by Kimball and later extensively developed by Henry

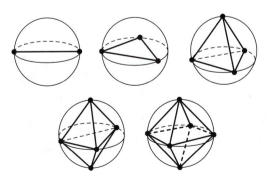

Figure 4.4 The points-on-a-sphere model. The most probable arrangements of two, three, four, five, and six points on the surface of a sphere that maximizes their distance apart.

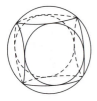

Figure 4.5 The octahedral arrangement of six equal circles on a sphere maximizes the area covered by the circles.

Bent in the 1960's. In this model, which Bent called the **tangent sphere model,** each electron pair domain is represented by a sphere. The arrangements adopted by the spheres are those that allow them to pack as closely as possible around a central spherical core. These sphere arrangements are easily demonstrated with Styrofoam spheres connected by elastic bands, or with toy balloons (Figures 4.6 and 4.7) They are found in nature in clusters of walnuts, for example (Figure 4.8). The intersection of the spheres with the spherical surface passing through the points of contact of the spheres gives the circles-on-a-sphere model, and the centers of the circles or the spheres give the points-on-a sphere model.

Attempts have been made to put the VSEPR model on a quantitative basis by describing the interaction between electron pairs in terms of force law of the type

$$U_{ij} = \frac{a}{d_{ij}^k}$$

where U_{ij} is the repulsion energy between the points i and j, d_{ij} is the distance between the two points, k is an integer, and a is a proportionality constant, to represent the interaction

Figure 4.6 Styrofoam sphere models representing the arrangements of two, three, four, five, and six valence shell electron pair domains.

Figure 4.7 Balloon models representing the arrangements of two to six valence shell electron pair domains.

Figure 4.8 Walnut clusters: an illustration from nature of the packing of spherelike objects. (Reproduced with permission from *In Our Image: Personal Symmetry in Discovery*, 1999, I. Hargittai and M. Hargattai, Kluwer, New York.)

between the points in a points-on-a-sphere model. The geometry that has the minimum total repulsion energy can then be calculated for any value of k. This model has been discussed in detail by Kepert. The difficulty with this model is that the appropriate value of k for any particular molecule is not known. The lower limit is $k = 1$, which unrealistically assumes only Coulombic repulsion between the domains considered as points. The geometry obtained as k approaches infinity is the same as that given by Bent's tangent sphere model. Fortunately for all except five and seven particles, the predicted arrangement is independent of the value of k. The arrangements for two to nine particles are summarized in Table 4.1. Another difficulty with this model is that if there is more than one type of ligand, an empirical parameter must to be introduced for each additional type of ligand as well as for a lone pair, to make allowance for the distortions of the geometry produced by the additional ligand types and lone pairs. Nevertheless, the model has proved useful, particularly for molecules containing bidentate and multidentate ligands. For this type of molecule Kepert found that $k = 6$ gave the best agreement with experiment.

Although the electron domain model is, as we shall see, a very useful model, we must remember that it is just that, a model—indeed a very approximate model. We cannot observe the individual domains of electrons but only the total electron density distribution.

In the next section we discuss the geometry of molecules that have two, three, four, and six valence shell electron pair domains for which all the foregoing models and all values of k predict the linear, triangular, tetrahedral, and octahedral arrangements of domains, respectively. The tetrahedron and the octahedron are two of the five regular polyhedra (Figure 4.9). A polyhedron is regular if its faces are equal polygons and each of its vertices has the same number of neighboring vertices in the same arrangement. The regular polyhedra have been known since classical times and are often called the Platonic solids because they played an important role in Plato's natural philosophy. The tetrahedron and the octahedron, which have equilateral triangular faces and also represent close-packed arrangements, are the two most important shapes in chemistry.

Table 4.1 Minimum Energy Geometries for n Repelling Particles Constrained to Move on the Surface of a Sphere

Number of particles	Arrangement[a]	Number of particles	Arrangement[a]
2	Linear	7	Monocapped octahedron[c,d]
3	Equilateral triangle	8	Square antiprism[d]
4	Tetrahedron	9	Tricapped trigonal
5	Trigonal bipyramid[b]		Prism[d]
6	Octahedron		

[a]For 2, 3, 4, 6, 8, and 9 particles, the minimum energy arrangements are independent of the value of k in the force law $U_{ij} = ad_{ij}^k$ describing the interaction between the points, where d_{ij} is the distance between the points i and j.

[b]For a value of $k = \infty$, which corresponds to a hard sphere model, the square pyramid has the same energy as the trigonal bipyramid.

[c]For $k = 1$, the pentagonal bipyramid has a lower energy than the monocapped octahedron.

[d]The geometries of these arrangements are described in Chapter 9.

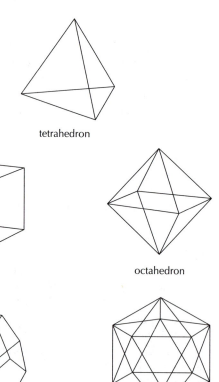

tetrahedron

cube

octahedron

dodecahedron

icosahedron

Figure 4.9 The five regular polyhedra.

We discuss molecules with a valence shell containing five electron pair domains in Section 4.6. The preferred arrangements of five valence shell domains, the trigonal bipyramid and the square pyramid, are not regular polyhedra and therefore exhibit special features not found in tetrahedral and octahedral molecules. Molecules with seven and more electron pair domains in the valence shell of a central atom are not common, although they are of considerable interest. They are restricted to the elements of period 4 and higher periods, with very small ligands such as fluorine, and are discussed in Chapter 9.

◆ 4.4 Two, Three, Four, and Six Electron Pair Domain Valence Shells

A useful terminology for classifying molecules according to their geometry is to denote the central atom by A, a singly bonded ligand by X, and an unshared electron pair by E. Thus H_2O, F_2O, SCl_2, and NH_2^- are all angular AX_2E_2 molecules; NH_3, NCl_3, PCl_3, and H_3O^+ are all triangular pyramidal AX_3E molecules; and CH_4, $SiCl_4$, BF_4^-, and NH_4^+ are all tetrahedral AX_4 molecules. Figure 4.10 shows tangent sphere models for AX_4, AX_3E, and AX_2E_2 molecules. Table 4.2 and Figure 4.11 summarize all the possible molecular shapes that result from valence shells containing two to six electron pair domains. The bond angles are predicted to be 180, 120, 109.5, or 90°, respectively. However, the bond angles in many molecules are only approximately equal to these ideal angles. Qualitative predictions of the deviations from these ideal bond angles can be made in many cases by taking into account differences in the sizes and shapes of the electron pair domains in a valence shell. We can think

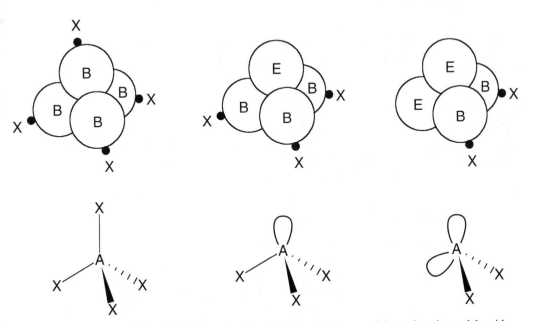

Figure 4.10 AX_4, AX_3E, and AX_2E_2 molecules: (a) tangent sphere models or domain models with spherical domains; B is a bonding pair and E is a lone pair and (b) conventional bond line structures.

Table 4.2 Electron Pair Arrangements and the Geometry of AX_nE_m Molecules[a]

Number of Electron Pairs	Arrangement of Electron Pairs	n	m	Class of Molecule	Shape of Molecule	Examples
2	Linear	2	0	AX_2	Linear	BeH_2, $BeCl_2$
3	Equilateral triangular	3	3	AX_3	Equilateral triangular	BCl_3, $AlCl_3$
		2	1	AX_2E	Angular	$SnCl_2$
4	Tetrahedral	4	0	AX_4	Tetrahedral	CH_4, $SiCl_4$
		3	1	AX_3E	Triangular pyramidal	NH_3, PCl_3
		2	2	AX_2E_2	Angular	H_2O, SCl_2
5	Trigonal bipyramidal	5	0	AX_5	Trigonal bipyramidal	PCl_5, AsF_5
		4	1	AX_4E	Disphenoidal	SF_4
		3	2	AX_3E_2	T-shaped	ClF_3
		2	3	AX_2E_3	Linear	XeF_2
6	Octahedral	6	0	AX_6	Octahedral	SF_6
		5	1	AX_5E	Square pyramidal	BrF_5
		4	2	AX_4E_2	Square planar	XeF_4

[a]n, number of bonding pairs; m, number of nonbonding pairs.

of an electron pair domain as a charge cloud of approximately constant volume but deformable in shape so that it has different shapes in different circumstances. In particular,

- Nonbonding or lone pairs have larger domains that occupy more angular space in the valence shell of the central atom than the domains of the bonding pairs.
- Bonding domains decrease in size and occupy less angular space around a central atom, with increasing electronegativity of the ligand and/or decreasing electronegativity of the central atom.
- Double-bond and triple-bond domains that consist of two and three electron pairs, respectively, are larger than single-bond domains.

4.4.1 The Effect of Lone Pairs on Bond Angles

Because a nonbonding pair is subject to the attraction of only one positively charged core, as opposed to two for a bonding pair, it is pulled in toward the core and its domain tends to surround the core as far as is permitted by the presence of the other valence electrons. In other words, it occupies more angular space. In contrast, a bonding pair is subject to the attraction of two cores and therefore has a smaller and more contracted domain that takes up less angular space around the core of A to an extent that depends on the electronegativity of X. In general, for a given atom A, the domains of nonbonding pairs take up more of the space around the core of A than the domains of bonding pairs. This is illustrated in Figure

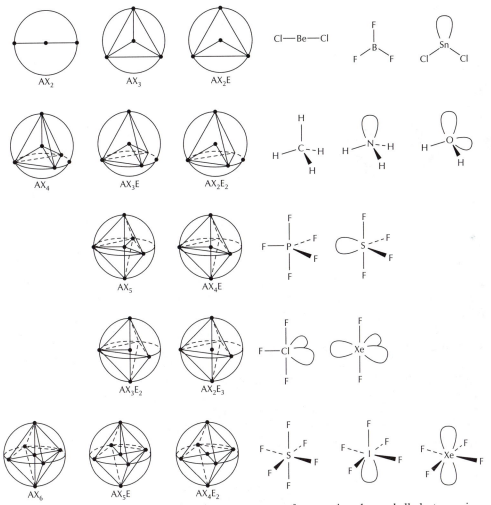

Figure 4.11 Molecular shapes based on the arrangements of two to six valence shell electron pairs.

4.12 for the tetrahedral arrangement of three bonding pairs and a lone pair in the NH_3 molecule, showing how this causes the bond angle to be smaller than 109.5°.

Because lone pair domains are larger and more spread out around the central core than bonding pair domains, the angles between bonding pair domains are smaller than the angles between lone pair domains. Consequently:

> In molecules with lone pairs, bond angles are smaller than the ideal values associated with a given number of equivalent pairs, and they decrease with increasing number of lone pairs.

Hence bond angles in AX_3E and AX_2E_2 molecules are generally smaller than the ideal angle of 109.5° as shown in Figure 4.12 and by the examples in Tables 4.3 and 4.4. However,

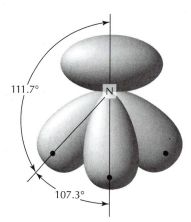

111.7°

107.3°

Figure 4.12 Representation of the bonding and nonbonding electron pair domains in the ammonia molecule, an AX_3E molecule.

for $(CH_3)_3N$, $(CH_3)_2O$, and OCl_2 in which oxygen or nitrogen is the central atom and has a greater electronegativity than the ligands, the bond angle is larger than the tetrahedral angle. In these molecules the oxygen and nitrogen valence shell electrons are not strongly localized into pairs and ligand–ligand repulsions are responsible for the larger than expected bond angles, as we will discuss in Chapter 5.

It should be noted that hydrogen, which like carbon is less electronegative than oxygen and nitrogen, nevertheless gives bond angles that are less than tetrahedral, suggesting that it localizes the electrons of the central atom to a greater extent than might be expected. That this is the case is confirmed by the analysis of electron density distribution, as we will see in Chapters 6 and 7. This unexpected behavior of hydrogen suggests that the assignment of a constant value for the electronegativity of hydrogen may not be as good an approximation as it seems to be for the other elements. Probably the ability of hydrogen to localize the electrons of the central atom is enhanced by the short lengths of bonds to hydrogen and its lack

Table 4.3 Bond Angles (°) in Some Trigonal Pyramidal AX_3E Molecules

Molecule	Bond Angle	Molecule	Bond Angle
NH_3	107.3	PBr_3	101.1
PH_3	93.8	$AsBr_3$	99.8
AsH_3	91.8	$SbBr_3$	98.2
SbH_3	91.7	PI_3	102
NF_3	102.2	AsI_3	100.2
PF_3	97.8	SbI_3	99.3
AsF_3	96.1	NMe_3	110.9
SbF_3	87.3	PMe_3	98.6
NCl_3	107.1	$AsMe_3$	96.0
PCl_3	100.3	$SbMe_3$	94.2
$AsCl_3$	98.6	SF_3^+	97.5
$SbCl_3$	97.2	SeF_3^+	94.5

Table 4.4 Bond Angles (°) in Some Angular
AX_2E_2 Molecules

Molecule	Bond Angle	Molecule	Bond Angle
H_2O	104.5	ClF_2^+	96
H_2S	92.3	BrF_2^+	92
H_2Se	90.6	ICl_2^+	93
H_2Te	90.3	HOF	97.3
OF_2	103.1	HOCl	102.5
SF_2	98.2	CF_3OF	104.8
SeF_2	94	CH_3OH	108.6
OCl_2	111.2	CH_3SH	96.5
SCl_2	102.8	CH_3SeH	95.5
$SeCl_2$	99.6	CH_3SCl	98.9
$TeCl_2$	97.0	CF_3SF	97.1
OMe_2	111.7	CF_2SCl	98.9
SMe_2	99.1	NH_2^-	99.4
$SeMe_2$	96.3	NF_2^-	96.7

of core electrons. Moreover, the small size of a hydrogen atom is another reason for the small HAH angles, as we discuss in Chapter 5.

Similarly, the presence of the lone pair in AX_5E molecules causes the bond angles to be less than 90° (Table 4.5). Moreover, as we see in Figure 4.13 the equatorial bonds A—X_{eq} that are adjacent to the lone pair are longer than the axial bond A—X_{ax}. The larger space requirement of the lone pair domain causes the four equatorial bond domains to move up toward the apex of the square pyramid, thus decreasing the $X_{ax}AX_{eq}$ bond angle, and it also causes them to move away from A, thus increasing the AX_{eq} bond lengths.

AX_4E_2 molecules always have a planar structure with two trans lone pairs and bond angles of 90° (Figure 4.14). The known examples of such molecules are ICl_4^-, ClF_4^-, BrF_4^-, IF_4^-, and XeF_4. This geometry allows the lone pair domains to occupy a maximum space and the bond angles to attain a maximum value of 90°.

Table 4.5 Bond Lengths and Bond Angles in Some Square Pyramidal
AX_5E Molecules

	Bond Lengths (pm)		
	Axial	**Equatorial**	**Bond Angle (°)**
$XeF_5^+(AgF_4^-)$	182.6	185.2	77.3
$XeF_5^+(PtF_6^-)$	181.0	184.3	79
ClF_5	157	167	86
BrF_5	168.9	177.4	84.8
IF_5	184.4	186.9	81.5
$TeF_5^-(Na^+)$	186.2	195.2	87.8
$SbF_5^{2-}(NH_4^+)_2$	191.6	207.5	88.0
$SbCl_5^{2-}(NH_4^+)_2$	236	258–269	85

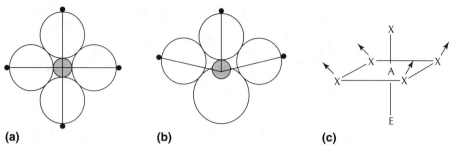

Figure 4.13 Section through an octahedral arrangement of six bonding electron pair domains, illustrating the effect of replacing one of these bonding domains by a larger nonbonding domain on bond lengths and bond angles (a) Four of the six equivalent bonding pairs in an octahedral arrangement. (b) Three of five bonding pairs and a nonbonding pair. The four bonding pairs adjacent to the lone pair are pushed up and away from the lone pair as also shown in (c), decreasing the bond angle and increasing the length of the bonds adjacent to the lone pair.

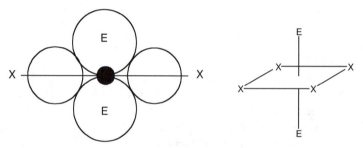

Figure 4.14 AX_4E_2 molecules have a planar structure with two trans lone pairs and bond angles of 90°. This geometry allows the lone pair domains to occupy a maximum of space in the valence shell and the bond angles to attain a maximum value of 90°.

The nonequivalence in the size and shape of bonding and nonbonding electron pair domains can alternatively be expressed in terms of the relative magnitude of their mutual Pauli repulsions, which decrease in the following order:

lone pair : lone pair > lone pair : bond pair > bond pair : bond pair

The VSEPR model was originally expressed in these terms, but because Pauli repulsions are not real forces and should not be confused with electrostatic forces, it is preferable to express the nonequivalence of electron pairs of different kinds in terms of the size and shape of their domains, as we have done in this chapter.

4.4.2 The Effect of Ligand Electronegativity on Bond Angles

When a ligand is more electronegative than the central atom, it draws the bonding electron density away from the central atom so that the space occupied by the bonding domain in the valence shell of the central atom decreases with increasing difference of electronegativity

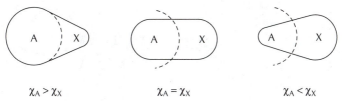

$$\chi_A > \chi_X \qquad\qquad \chi_A = \chi_X \qquad\qquad \chi_A < \chi_X$$

Figure 4.15 The size of a bonding pair domain in the valence shell of A decreases with increasing electronegativity of X.

between the ligand and the central atom (Figure 4.15). Consequently, in molecules in which there are large lone pair domains on the central atom, the bonding domains are pushed more closely together the greater the electronegativity of the ligand, so that

> bond angles decrease with increasing electronegativity of the ligand or decreasing electronegativity of the central atom.

Many examples of this effect can be found in Tables 4.3, 4.4, and 4.5. For example, the bond angle decreases from 102° in PI_3 to 97.8° in PF_3 as χ_X increases, and from 102.3° in NF_3 to 87.3° in SbF_3 as χ_A decreases. We note also that in a series of group 5, or group 6 molecules such as the chlorides of group 5 the largest decrease in the bond angle occurs from NCl_3 (107.1°) to PCl_3 (100.3°), with only a much smaller decrease to $SbCl_3$ (97.2°) consistent with the large decrease in electronegativity from N (3.1) to P (2.1) and the much smaller subsequent decrease to Sb (1.8).

It is also significant that the bond angles in period 2 molecules are not much smaller than the tetrahedral angle, whereas those for molecules of the later periods approach values of 90° and in a few cases are even smaller. This observation is consistent with period 2 molecules being able to accommodate only four electron pairs in their valence shell, whereas the larger valence shells of the elements of period 3 and later periods may accommodate six or even more electron pairs. This has important consequences for molecules of an element from period 3 and beyond in which there are only four electron pair domains in the valence shell, one of which is a lone pair. In such a molecule the bonding domains are easily pushed together by the lone pair until the bond angle approaches the limiting value of 90° found in the octahedral arrangement of six domains. So the bond angles in period 3 molecules that have only four domains in the valence shell are generally much smaller than in the corresponding period 2 molecule. In general it may be seen in Tables 4.3, 4.4, and 4.5 that bond angles decrease with increasing size of the central atom as more space becomes available for the electron pair domains, allowing the bonding domains to be pushed together more easily, thus decreasing bond angles.

◆ 4.5 Multiple Bonds

In ethene each carbon atom has four electron pairs in its valence shell, which may be considered to occupy four domains with an approximately tetrahedral arrangement. Two of these pairs are forming the double bond, so that the two tetrahedra are sharing an edge, giving an

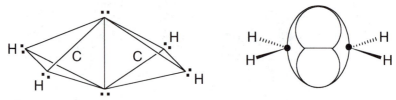

Figure 4.16 Double bond: (a) Lewis model of two tetrahedra sharing an edge. (b) Domain model: the two single electron pair domains of the double bond are pulled in toward each other by the attraction of the two carbon cores forming one four-electron double-bond domain with a prolate ellipsoidal shape, thereby allowing the two hydrogen ligands to move apart.

overall planar geometry (Figure 4.16a). This model is equivalent to the bent-bond model (Figure 1.4). According to this model the double bond consists of two adjacent electron pair domains, one on each side of the internuclear axis (Figure 4.16b). Because they lie off the internuclear axis, the two domains of the double bond are attracted toward the internuclear axis by both the carbon atoms. Thus the angle between the two domains is reduced to less than 109° at each of the carbon atoms. The reduction in this angle increases the angle that each of these two domains makes with each of the CH bonding domains, reducing the repulsion on these domains and thus allowing them to separate to an angle larger than 109°. Table 4.6 gives values for XCX bond angles in some $X_2C{=}CY_2$ molecules. In all cases the bond angles are larger than 109°, consistent with this model.

Table 4.6 Bond Angles in Some $X_2C[dbond]CY_2$ Molecules

Molecule	Bond	Bond Length (pm)	Bond Angles (°)			
			X—C—X	Y—C—Y	X—C=C	Y—C=C
$H_2C{=}CH_2$	C—H	108.7	117.4	117.4	121.3	121.3
	C=C	133.9				
$F_2C{=}CH_2$	C—F	131.5	110.6	119.3	124.7	120.3
	C—H	109.1				
	C=C	134.0				
$F_2C{=}CF_2$	C—F	131.9	112.4	112.4	123.8	123.8
	C=C	131.3				
$F_2C{=}CCl_2$	C—F	131.5	112.1	119.6	124.0	120.5
	C—Cl	170.6				
	C=C	134.5				
$Cl_2C{=}CCl_2$	C—Cl	171.9	115.6	115.6	122.2	122.2
	C=C	135.5				
$Br_2C{=}CBr_2$	C—Br	188.2	115.2	115.2	122.4	122.4
	C=C	136.3				
$I_2C{=}CI_2$	C—I	210.6	114.2	114.2	122.9	122.9
	C=C	136.3				
Me_2CvCMe_2	C—C	151.1	113.2	113.2	123.4	123.4
	C=C	135.1				
$Me_2C{=}CH_2$	C—C	150.7	115.6	117.4	122.2	121.3
	C—H	109.5				

The bent-bond model for the double bond has sometimes been criticized because it appears to suggest that there is no electron density along the internuclear axis. However, we should beware of interpreting such a bond diagram too literally, because each electron pair domain occupies a considerable volume of space and the domains overlap. The two domains are attracted toward the internuclear axis (i.e., toward each other) by the two carbon cores (Figure 4.16). Consequently, they overlap so much that the total electron density resulting from the two domains is a maximum along the internuclear axis, as we will see in Chapter 6. Thus instead of using two adjacent electron pair domains to represent the double bond, it can alternatively be represented by a single four-electron domain in which the two electron pairs cannot be distinguished. This domain has a greater extent in the direction perpendicular to the molecular plane than in the molecular plane, that is, it has an ellipsoidal cross section (Figure 4.16b). This expected shape is consistent with the electron density distribution in a double bond, discussed later in Chapter 6.

In the classical model of ethyne, one of the four tetrahedrally arranged electron pairs in the valence shell of each carbon atom is used to form the CH bond and the remaining three form the triple bond. In other words, the two tetrahedra are sharing a face so that the molecule has an overall linear geometry as shown in Figure 4.17a. The corresponding domain model would represent the triple bond by three adjacent electron pair domains, as shown in Figure 4.17b. However, this is not a very realistic model because in a linear molecule the multiple-bond electrons do not form pairs, just as we have seen for the nonbonding electrons in HF (Figure 4.3). The two tetrahedra of opposite-spin electrons around each carbon in ethyne are not brought into coincidence, except in the CH bonds, so that to minimize their repulsion energy, the six electrons of the triple bond have a most probable distribution in which they are equally distributed in a ring surrounding the CC axis (Figure 4.18). This is the description of the molecule given by Linnett's double-quartet model (Box 4.1). The bonding electrons have an equal probability of being found anywhere around this ring, giving an overall electron density distribution with a circular cross section, which as we will see in Chapter 6 is confirmed by the calculated electron density distribution. The classical description of a triple bond in a linear molecule is therefore misleading, while that given by

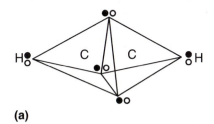

(a)

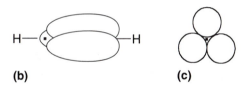

(b)　　　　　**(c)**

Figure 4.17 Triple bonds: (a) Lewis model of two tetrahedra sharing a face, (b) three electron pair domains, and (c) end-on view of the three electron pair domains forming the triple bond.

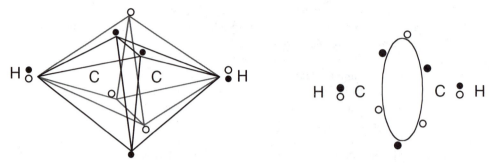

Figure 4.18 Model of ethyne showing that only the electrons of the C—H bonds are localized into pairs. The remaining six electrons are evenly distributed in a ring around the C—C axis.

▲ BOX 4.1 ▼
The Double-Quartet Model

Linnett used the concept that an octet of valence shell electrons consists of two sets of four opposite-spin electrons to show that in diatomic and other linear molecules the two tetrahedra are not in general formed into four pairs as we have discussed for F_2 and the CC triple bond in C_2H_2. This idea is the basis of the double-quartet model, which Linnett applied to describe the bonding in a variety of molecules. It is particularly useful for the description of the bonding in radicals, including in particular the oxygen molecule, which has two unpaired electrons and is therefore paramagnetic This unusual property is not explained by the Lewis structure

$$:\ddot{O}::\ddot{O}:$$

which assumes that all 12 electrons are paired and therefore predicts a diamagnetic molecule. However, because oxygen is a diradical there must be seven electrons of α spin and five of β spin. The set of seven α electrons will be at the vertices of two tetrahedra that share one vertex while the set of five β electrons will be at the vertices of two tetrahedra sharing a face.

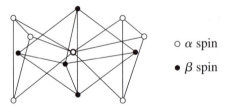

○ α spin

● β spin

This gives an overall arrangement in which none of the electrons are paired and the electron repulsion energy is less than in the Lewis structure, hence is the preferred arrangement. All together there are four bonding electrons—three of α spin and one of β spin—so that there is effectively a double bond as in the Lewis structure.

The bonding in a diatomic molecule with a single unpaired electron such as NO can be described in an analogous manner. In this case there are six electrons of α spin and five of β spin, with the following arrangement.

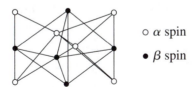

○ α spin

● β spin

Thus there are five bonding electrons giving a bond order of 2.5, consistent with the bond length of 115 pm, versus 121 pm for the four-electron bond in O_2 and 110 pm for the six-electron bond in N_2. For these and other related molecules, the double-quartet model is a convenient and useful alternative to the conventional molecular orbital model. Moreover, it shows that for a singly bonded terminal atom such as F or Cl there is a ring of six nonbonding electrons rather than three separate lone pairs. As we will see in Chapters 7 and 8, this conclusion is confirmed by the analysis of electron density distributions.

the double-quartet model is a better approximation. Linear molecules are an exception to the rule that electrons are found in pairs in molecules with an even number of electrons.

We saw in Chapter 2 (Table 2.9) that a CC double bond is not twice as strong as two single CC bonds and a triple bond not three times as strong as three single CC bonds. The relative weakness of CC double and triple bonds can be attributed to the increased repulsion between the electrons when they are drawn closer together in a multiple-bond domain. If one of the bonds in a double bond is broken, the remaining bond is stronger than half the double-bond strength because the repulsion between the two electron pairs in the double bond is no longer present.

The formation of a multiple bond requires that two or three pairs of electrons be attracted into the same bonding region. This implies that these electrons must be sufficiently strongly attracted to overcome their mutual repulsion. Thus strong multiple bonds are formed only between the most electronegative atoms, in particular carbon, nitrogen, and oxygen. Fluorine, having only a single unpaired electron, is limited to the formation of one single bond. Other electronegative elements such as the elements in groups 15–18 in their higher oxidation states form strong multiple bonds with oxygen, nitrogen, and carbon (Figure 4.19). These multiple bonds differ from those between carbon, nitrogen, or oxygen atoms in that they are very polar, which increases their strength.

The elements in groups 15–18 in period 3 and beyond also form multiple bonds *with each other*. But because these elements have smaller electronegativities than the corresponding elements of period 2, they are appreciably weaker than the multiple bonds between

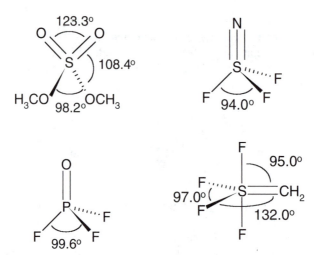

Figure 4.19 Molecules of phosphorus and sulfur with multiple bonds to carbon, nitrogen, and oxygen.

period 2 elements, as shown by the mean bond enthalpies in Table 2.8. Double bonds between the atoms of the elements of period 3 and beyond are also much more reactive than period 2 double bonds. This is because the bonding electrons are less strongly held and because the valence shells of these elements are larger and less crowded than the valence shells of period 2 elements and are therefore more susceptible to attack. Consequently these bonds are generally found in stable molecules only when access to the multiple bond is blocked by very large ligands.

We can use the concept of double- and triple-bond domains as the basis for a simple and convenient method for predicting the shapes of molecules containing double and triple bonds. Instead of considering the arrangement of a given number of single electron pair domains in a valence shell, we consider the arrangement of the total number of domains (lone pair, single-bond, double-bond, or triple-bond domains). According to this model,

> the arrangement of the bonds around an atom depends only on the total number of domains (lone pair, single bond, double bond, or triple bond) in the valence shell and is independent of the bond order, that is, whether the bonds are single, double, or triple, or of an intermediate character.

Figure 4.20 summarizes the predicted shapes, and gives examples, of molecules containing multiple bonds. *In each case the predicted geometry is in agreement with experiment.* This model is convenient for molecules with multiple bonds because it readily accommodates molecules in which the bonds have only a partial multiple bond character, such as benzene and the carbonate ion, in which the bonding is often described in terms of resonance structures (Chapter 2). For example, each carbon atom in benzene has three bonding domains in its valence shell, as does the carbon atom in the carbonate ion, and in both cases these domains have a planar triangular arrangement around the carbon atom.

Total Number of Domains	Arrangement	Bonding Domains	Lone Pair Domains	Geometry	Examples
2	Linear	2	0	Linear	$O=C=O$ $H-C\equiv N$
3	Triangular	3	0	Triangular	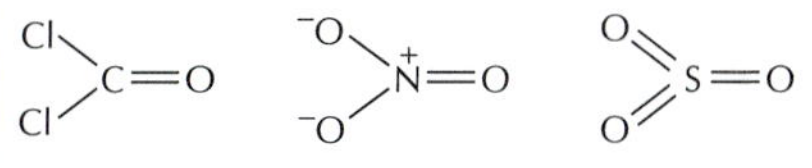 $Cl_2C=O$, $^-O_2N^+=O$, $O_2S=O$
		2	1	V-Shape	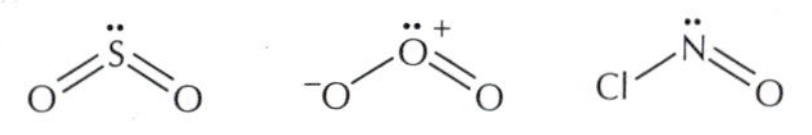 O_2S (lone pair), $^-O\,\ddot{O}^+O$, $Cl-N\ddot{}=O$
4	Tetrahedral	4	0	Tetrahedral	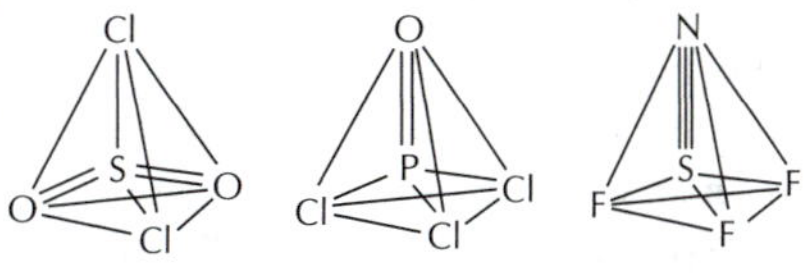 O_2SCl_2, $OPCl_3$, SF_4 (with O)
		3	1	Trigonal Pyramid	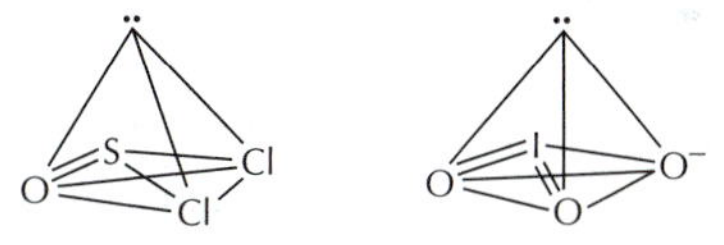 O_2SCl_2 (lone pair), IO_3^-
		2	2	V-Shape	$O_2Cl O^-$, O_2Xe
5	Trigonal Bipyramid	5	0	Trigonal Bipyramid	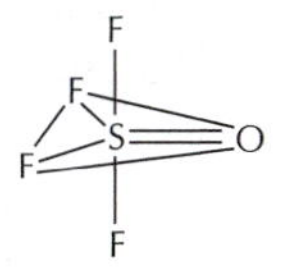 SF_4O
		4	1	Irregular Tetrahedron	 O_2XeF_2 :, $O_2IF_2^-$:
6	Octahedron	6	0	Octahedron	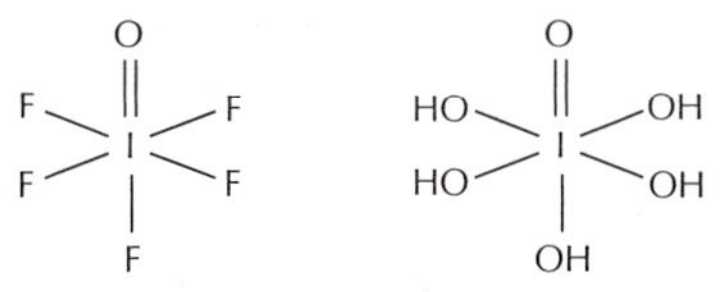 IOF_5, $I(OH)_5O$

Figure 4.20 Predicted shapes of molecules containing multiple bonds.

Bond angles in molecules containing multiple bonds deviate from the ideal values because a double-bond domain is larger than a single-bond domain, and a triple-bond domain is even larger. Thus

angles involving triple bonds are larger than those involving double bonds, which in turn are larger than those between single bonds.

The structure of dimethyl sulfate, $(CH_3O)_2SO_2$, provides a good examples because it has three types of OSO bond angle that decrease in magnitude in the expected order (Figure 4.19), namely double bond–double bond > double bond–single bond > single bond–single bond. Further examples are given in Figure 4.19. In most of these cases the single and double bonds are to the atoms of different elements, but nevertheless the angles are in the sequence given above.

4.5.1 Comparison of the Domain and Orbital Models of Multiple Bonds

Since the advent of orbital models, the bent-bond model has been largely superseded by the σ–π model. The σ–π model is more useful for delocalized systems such as aromatic molecules. For simple molecules such as ethene, however, the bent-bond model is just as useful and indeed has some advantages over the σ–π model. For example, it *predicts* the planar geometry of the ethene molecule, whereas the σ–π model does not. Indeed, we can use the σ–π description of the bonding only when a molecule has a plane of symmetry through the double bond. On the basis of the known planar geometry around each carbon, the σ–π model *assumes* that the three σ bonds are formed from sp^2 hybrid orbitals. Then making the remaining singly occupied p orbitals on each carbon parallel to give maximum overlap gives the molecule an overall planar geometry, as discussed in Chapter 3 (Figure 3.18).

In its simplest form, the bent-bond model predicts an XCX angle of 109° in ethene and substituted ethenes. The σ–π model is based on the assumption that the carbon atoms are sp^2-hybridized and that the XCX bond angle is therefore 120°. In most molecules the observed angles have values intermediate between these two values (Table 4.6). We can understand this in terms of the domain model because each of the two domains forming the double bond is attracted toward the CC axis, reducing the angle between them, thus allowing more space for the CX bonding domains. Consequently, the XCX bond angle increases to a value larger than 109.5°. Alternatively, if we consider the double bond to be formed by one four-electron domain so that each carbon atom has two single-bond domains and a double-bond domain, we predict an angle between the single-bonds domains of somewhat less than 120° because of the larger size of the double bond domain. Using the hybrid orbital model, we can adjust the degree of hybridization between the limits of sp^2 and sp^3 to match the observed bond angle, but we cannot predict the angle.

The bent-bond model can be expressed in orbital terms by assuming that the two components of the double bond are formed from sp^3 hybrids on the carbon atoms (Figure 3.19) That this model and the σ–π model are alternative and approximate, but equivalent, descriptions of the same total electron density distribution can be shown by converting one into the other by taking linear combinations of the orbitals, as shown in Figure 3.20. But neither form of the orbital model can predict the observed deviations from the ideal angles of 109° and 120°.

◆ 4.6 Five Electron Pair Valence Shells

Five points can be arranged on the surface of a sphere such that they are all equivalent, only in a planar pentagonal arrangement, which does not maximize the distance between the points. In other words, there is no regular polyhedron with five equivalent vertices. There are two

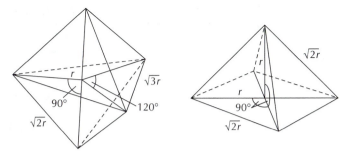

Figure 4.21 Geometry of the trigonal bipyramidal and the square pyramidal arrangements of five points on the surface of a sphere.

arrangements of five points that maximize the shortest distance between any pair of points, the trigonal bipyramid and the square pyramid (Figure 4.21). In these two arrangements the shortest distance between any two points is $\sqrt{2}r$, where r is the radius of the sphere. However, the trigonal bipyramid has only six such shortest distances, whereas the square pyramid has eight. On this basis, therefore, the trigonal bipyramid is the preferred geometry.

Similarly, five spheres (representing either bonding domains or lone pair domains) can be packed around a spherical central core so that they are all equivalent only in the planar pentagonal arrangement, which does not, however, allow them to get as close as possible to the central core. The two geometries that allow them to get as close as possible to the central core are the trigonal bipyramid and the square pyramid (Figure 4.21). In the square pyramid geometry, each sphere in the base is touching three other spheres and the sphere at the apex is touching four others, giving total of eight contacts. In the trigonal bipyramid geometry the axial spheres are touching all three equatorial spheres but the equatorial spheres are not touching each other, giving a total of only six contacts. On the basis of minimizing the number of contacts, the trigonal bipyramidal geometry is preferred. With very few exceptions, the geometry of AX_5, AX_4E, AX_3E_2, and AX_3E_3 molecules is based on the trigonal bipyramidal arrangement of five domains. Some examples are given in Table 4.7.

Unlike the tetrahedron and the octahedron, which are regular solids with four and six equivalent vertices, respectively, the trigonal bipyramid has two sets of vertices that are not equivalent. The axial vertices have three close neighbors, whereas the equatorial vertices have two close neighbors (the axial vertices) and two more that are farther away (the other two equatorial vertices). This nonequivalence of the vertices of a trigonal bipyramid has several important consequences

As we can see in Table 4.6:

The axial bonds in a trigonal bipyramidal molecule are longer than the equatorial bonds.

When five repelling points are situated on the surface of a sphere, they are not at true equilibrium because the axial points suffer a greater repulsion from their neighbors than the equatorial points. Each axial point has three close neighbors at 90°, while an equatorial point has only two close neighbors at 90° (the two axial points) and two neighbors further away at 120°. Given that the repulsive force falls off very rapidly with distance, there is a greater to-

Table 4.7 Bond Lengths and Bond Angles in Some AX_nE_{5-n} Molecules

Molecule	Bond Lengths (pm)		Bond angles (°)	
	Axial	Equatorial	$X_{ax}AX_{ax}$	$X_{eq}AX_{eq}$
AX_5				
PCl_5	214	202	180	120
PF_5	157.7	153.4	180	120
$SbCl_5$	243	231	180	120
AsF_5	171.1	165.6	180	120
$Sb(CH_3)_5$	220.9	213.5	180	120
AX_4E				
PF_4^-	174	160	168.3	99.9
SF_4	164.6	154.5	167.0	101.6
SeF_4	177.1	168.2	169.2	100.6
IF_4^+	184	177	160	92
AX_3E_2				
ClF_3	169.8	159.8	175	—
BrF_3	181.0	172.1	172.4	—
XeF_3^+	190.5	183.5	160.9	—
AX_2E_3				
XeF_2	197.7	—	180	—
ICl_2^-	255	—	180	—

tal repulsive force on the axial points than on the equatorial points. Consequently, if the re-straining spherical surface is removed and replaced by a force attracting the points toward the center of the sphere, the axial points will move away from the center of the sphere, allowing the equatorial points to move closer until equilibrium is attained. Thus in a trigonal bipyramidal molecule the axial bonds are longer than the equatorial bonds. This difference in bond lengths is simply a consequence of the geometry of the trigonal bipyramid and, as we will see in Chapter 9, it cannot be explained by any orbital model.

Because the axial positions have more close neighbors than the equatorial positions, there is more space available to a ligand or its bonding domain in an equatorial position than in an axial position. Thus nonbonding domains and larger bonding domains preferentially occupy the equatorial positions. Consequently

in AX_5, AX_4E, and AX_3E_2 molecules in which there are ligands with different electronegativities, the less electronegative ligands preferentially occupy the equatorial positions and the more electronegative ligands preferentially occupy the axial positions.

Some examples are given in Figure 4.22.

Because there is more space available in the equatorial than in the axial positions,

the lone pair domains always occupy the equatorial positions in $AX_{5-n}E_n$ molecules.

Thus all AX_4E molecules have the shape of a trigonal bipyramid with one missing vertex, which can be formally called a disphenoid. It is sometimes described as the SF_4 geometry

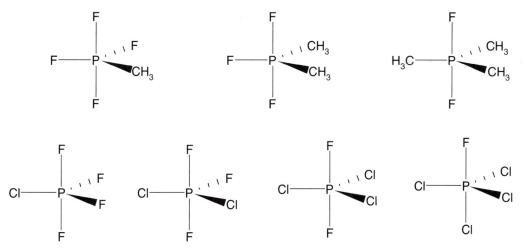

Figure 4.22 The less electronegative ligands always preferentially occupy the less crowded equatorial sites of a trigonal bipyramid, leaving the more electronegative ligands in the more crowded axial sites.

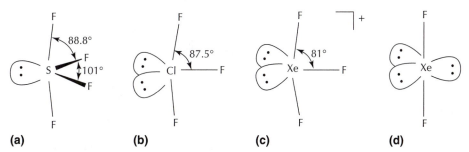

Figure 4.23 Lone pairs always occupy the less crowded equatorial sites of a trigonal bipyramid. (a) The disphenoidal geometry of the AX_4E molecule SF_4. (b) and (c) T-shaped AX_3E_2 molecules. (d) A linear AX_2E_3.

(Figure 4.23). or as the "sawhorse" geometry. All AX_3E_2 molecules have a T shape, and all AX_2E_3 molecules are linear (Figure 4.23). In all cases, because of the larger size of a lone pair domain, the equatorial–equatorial bond angle is smaller than 120° and the axial–equatorial bond angle is smaller than 90°.

With increasing size of the central atom, the lone pair domain is able to take up more space in the valence shell of the central atom, pushing the ligands close together. So the bond angles decrease with increasing size of the central atom just as we saw for AX_3E and AX_2E_2 molecules. As we see by the examples in Figure 4.24, multiple bond domains occupy more of the angular space around a core,

multiple bond domains always occupy the less crowded equatorial sites in an AX_5 molecule.

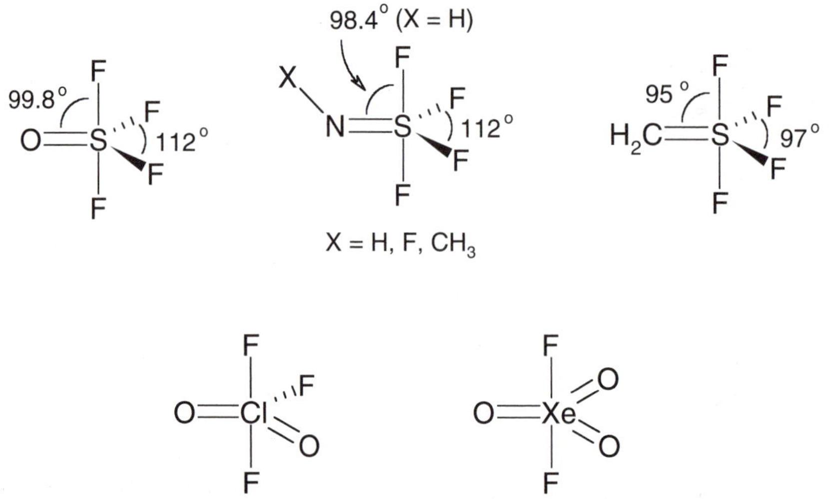

Figure 4.24 Double bonds always occupy the less crowded equatorial sites of a trigonal bipyramid and form larger bond angles than single bonds.

Figure 4.24 also shows that multiple bonds distort the bond angles in the same way as a lone pair.

◆ 4.7 Limitations and Exceptions

There are a number of apparent exceptions to the VSEPR model, most of which can be classified into two main groups: those due to ligand–ligand interactions and those due to the nonspherical cores found in transition metal molecules.

4.7.1 Ligand–Ligand Interactions

We have already mentioned that the bond angles in $N(CH_3)_3$, $O(CH_3)_2$, and OCl_2 are larger than $109.5°$. A similar exception is the Li_2O molecule, which according to ab initio calculations is linear. As we shall see in Chapters 7 and 8, these exceptions arise because in these molecules the ligands are not electronegative enough to localize the electrons in the valence shell of the central atom into pairs. Therefore the VSEPR model is not valid for these molecules. In the case of Li_2O, the electrons in the valence shell of the central oxygen atom are so poorly localized that the central oxygen is close to being a free oxide ion with a spherical electron density distribution. In such a case, the geometry is dominated by ligand–ligand interaction, which is minimized in the observed linear geometry. In Chapter 5 we will see that ligand–ligand interactions are also important in many molecules to which the VSEPR model applies. Consideration of these interactions gives us a still better understanding of the geometry of these molecules.

The VSEPR model emphasizes the geometric arrangement of bonding and nonbonding electron pairs as the major factor determining molecular geometry. The possibility that repulsive interactions between the ligands might play an important role in determining the geometry of molecules has also been considered from time to time. These interactions have generally been regarded as relatively unimportant except for molecules with very bulky ligands such as the *t*-butyl group. Recently, however, it has been shown that such interactions may be much more important than has generally been supposed, and a new model of molecular geometry, the **ligand-close packing (LCP) model,** has been developed based on ligand–ligand repulsions. The LCP model is applicable to many molecules, but it is particularly useful for molecules with small central atoms such as those of period 2, and for molecules in which the ligands are insufficiently electronegative to strongly localize the electrons in the valence shell of the central atom into pairs. It provides good explanations for many of the exceptions to the VSEPR model, and it can give more quantitative predictions of the deviations of bond angles from the ideal angles of 90, 109.5, and 120° than the VSEPR model. We discuss the LCP model in Chapter 5.

4.7.2 Transition Metal Molecules

The core of a transition metal in a molecule is generally not spherical because unlike the core of a main group element, it does not consist of completed shells. The outer shell of the core is often incomplete and contains nonbonding electrons. For example, in the series of molecules $ScCl_2$, $TiCl_2$, VCl_2, $CrCl_2$, $MnCl_2$, . . . , the transition metal has respectively 1, 2, 3, 4, and 5, . . . , nonbonding electrons. These nonbonding electrons are found in the outer shell of the core rather than in the valence shell, as in a molecule of a main group element. As we have seen, nonbonding or lone pair electrons in a valence shell have an important effect on bond angles and lengths, which is described by the VSEPR model. In contrast, the nonbonding electrons in the core of a transition metal have a smaller and less direct effect on molecular geometry. Because of the presence of these nonbonding electrons, the core generally does not have the spherical shape characteristic of the core of main group elements. This nonspherical shape of the core affects the geometry of the molecule in a different manner from that of nonbonding electrons in the valence shell of a nonmetal atom. We discuss the geometry of transition metal molecules again in Chapter 9.

▶ Further Reading

R. J. Gillespie, *Molecular Geometry,* 1972, Van Nostrand Reinhold, London.

R. J. Gillespie and I. Hargittai, *The VSEPR Model of Molecular Geometry,* 1991, Allyn & Bacon.

D. W. Kepert, *Inorganic Stereochemistry*, 1986, Springer-Verlag, New York.
    A detailed discussion of geometry in terms of the points-on-a-sphere model.

I. Hargittai, *The Structure of Volatile Sulphur Compounds,* 1985, Reidel, Dordrecht.
    Detailed discussion of molecules with a central sulfur atom with many applications of the VSEPR model.

I. Hargittai and M. Hargittai, *Symmetry Through the Eyes of a Chemist.* 1st ed., 1986, 2nd ed., 1998, VCH, New York.

An interesting book that contains a good summary of the VSEPR model.

J. W. Linnett, *The Electronic Structure of Molecules,* 1964, Methuen, London.

Covers the double-quartet model in detail.

Recent reviews of the VSEPR model and its applications.

R. J. Gillespie and E. A. Robinson, *Angew. Chem. Int. Ed. Engl. 35,* 495, 1996.

R. J. Gillespie, *Chem. Soc. Rev., 52,* 1992.

R. J. Gillespie, *J. Chem. Educ. 69,* 742, 1992.

R. J. Gillespie, *Coord. Chem. Rev. 197,* 51, 2000.

▶ References

H. A. Bent, *J. Chem. Educ. 40,* 446, 523, 1963; **42,** 302, 348, 1965; **44,** 512, 1967; **45,** 768, 1968.

R. J. Gillespie and R. S. Nyholm, *Q. Rev. Chem. Soc. 11,* 339, 1957.

O. Knop, E. M. Palmer, and R. W. Robinson, *Acta Crystallogr. A31,* 19, 1975.

The paper gives more information on the points-on-a-sphere model.

N. V. Sidgwick and H. E. Powell, *Proc. R. Soc. London A176,* 153, 1940.

C H A P T E R

5

LIGAND–LIGAND INTERACTIONS
AND THE LIGAND
CLOSE-PACKING (LCP) MODEL

◼ ◼ ◼

◆ 5.1   Introduction

The VSEPR model assumes that the geometry of a molecule is determined by the arrangements of bonding and nonbonding electron pairs in the valence shell of an atom, and it may
therefore be called an *electronic model.* In this chapter we consider an apparently different
model that assumes that the geometry of a molecule is determined simply by the repulsive
interactions between the ligands and may therefore be called a *steric model.* According to
this model, the geometry of an AX_n molecule is the geometry that allows n ligands to approach as closely as possible to the central atom A, that is, to adopt a close-packed arrangement around A, thus minimizing the energy of the molecule. This model is called the **lig-
and close-packing (LCP) model.**

The close packing of anions around cations has long been recognized as one of the main
factors determining the structures of ionic crystals. Similarly, the structures of many molecular solids are determined by the tendency of the molecules to pack together as closely as
possible under the influence of the relatively weak van der Waals attractive forces that act
between all closed-shell atoms and molecules. When the molecules reach their equilibrium
positions, this attractive force is just balanced by the repulsive "Pauli force" that arises from
the distortion of their electron distributions as they resist overlapping each other. We saw a
simple example of a "Pauli force" in Chapter 3. The repulsion between two helium atoms
resulting from the distortion of their charge clouds as they resist overlapping as the two He
atoms come together prevents the formation of a stable molecule.

To a first approximation, atoms in molecules may be regarded as hard spheres with a
segment cut off in the bonding direction, as in the familiar space-filling models. The radius
of the atom in a nonbonding direction is called the **van der Waals radius.** Half the distance
between two atoms of the same kind in adjacent molecules at equilibrium is taken as the van
der Waals radius (Figure 5.1). In assigning a fixed radius in this way, we assume that atoms

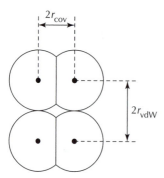

Figure 5.1 Van der Waals and covalent radii in a diatomic molecule.

have a spherical shape except in a nonbonding direction, which may not be correct, and that they are incompressible, which is only very approximately true. The van der Waals radius of an atom does not have a truly constant value because it depends to some extent on the strength of the forces holding the molecules together. Moreover, atoms in molecules are not truly spherical, so that they may have a different nonbonded radius in different directions, and they may also be more compressible in one direction than another. Figure 5.2a shows the electron density distribution of the Cl_2 molecule, a topic we discuss in more detail in

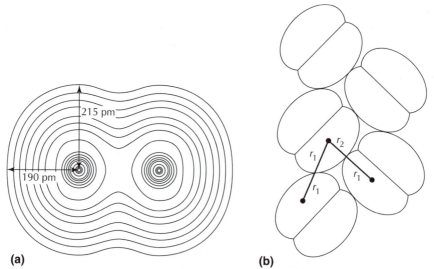

(a)   (b)

Figure 5.2 (a) Electron density contour map of the Cl_2 molecule (see Chapter 6) showing that the chlorine atoms in a Cl_2 molecule are not portions of spheres; rather, the atoms are slightly flattened at the ends of the molecule. So the molecule has two van der Waals radii; a smaller van der Waals radius, $r_2 = 190$ pm, in the direction of the bond axis and a larger radius, $r_1 = 215$ pm, in the perpendicular direction. (b) Portion of the crystal structure of solid chlorine showing the packing of Cl_2 molecules in the (100) plane. In the solid the two contact distances $r_1 + r_1$ and $r_1 + r_2$ have the values 342 pm and 328 pm, so the two radii are $r_1 = 171$ pm and $r_2 = 157$, pm which are appreciably smaller than the radii for the free Cl_2 molecule showing that the molecule is compressed by the intermolecular forces in the solid state.

Chapter 6. The outer contour in this map is for a density of 0.001 au, which has been found to represent fairly well the outer surface of a free molecule in the gas phase, giving a value of 190 pm for the radius in the direction opposite the bond and 215 pm in the perpendicular direction. In the solid state molecules are squashed together by intermolecular forces giving smaller van der Waals radii. Figure 5.2b shows a diagram of the packing of the Cl_2 molecules in one layer of the solid state structure of chlorine. From the intermolecular distances in the direction opposite the bond direction and perpendicular to this direction we can derive values of 157 pm and 171 pm for the two radii of a chlorine atom in the Cl_2 molecule in the solid state. These values are much smaller than the values for the free molecule in the gas phase. Clearly the Cl_2 molecule is substantially compressed in the solid state. This example show clearly that the van der Waals of an atom radius is not a well defined concept because, as we have stated, atoms in molecules are not spherical and are also compressible.

Nevertheless values for the van der Waals radii of atoms are often quoted in textbooks and some typical values are given in Table 5.1. Their accuracy is no more than ± 5 pm. We see that the radius usually given for chlorine (180 pm) is substantially larger than the two values given above for the Cl atom in the Cl_2 molecule in the solid state. This discrepancy arises because chlorine has a negative charge in most molecules other than Cl_2 (a consequence of its large electronegativity). As we discuss later in this chapter, the larger the negative charge of an atom the greater is its size. Indeed the often quoted value of 180 pm for the van der Waals radius of the chlorine atom is essentially the same as its ionic radius of 181 pm (Tables 2.4 and 2.6), implying that in most molecules a chlorine atom has a large negative charge, as we will see later is indeed the case. Radii of the isolated free atoms obtained from calculated electron densities (Chapter 6) are also given in Table 5.1. The radii of isolated free atoms are larger than the van der Waals radii obtained for the solid state by approximately 10 pm but they vary in the same way as the van der Waals radii.

Van der waals radii have been used in several different ways, although any conclusions drawn from their use must be viewed with appropriate skepticism because of their approximate nature. For example, when two atoms in different molecules have an interatomic dis-

Table 5.1 Van der Waals, 1,3 Nonbonded Radii and Radii of Isolated Free Atoms (pm)

Element	1,3 Nonbonded Radius	Van der Waals Radius	Radius of Isolated Gas Phase Atom[a]
H	92[b]	120	134
C	125[b]	—	173
N	114[b]	150	162
O	113[b]	140	154
F	108[b]	135	147
Si	155[c]	—	212
P	145[c]	190	203
S	145[c]	185	197
Cl	144[b]	180	189

[a]From calculated electron densities (Chapter 6).

[b]1,3 radii due to Bartell (1960).

[c]1,3 radii due to Glidewell (1975, 76).

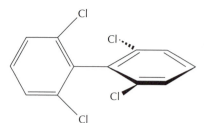

Figure 5.3 Steric interaction between the ortho chlorine atoms in biphenyl prevents the two rings from adopting a coplanar orientation.

tance that is substantially shorter than the sum of their van der Waals radii it is often considered that there is a bonding interaction between them. They have also been used to account for an otherwise unexpected geometry of a molecule. For example, two chlorine substituents in ortho positions in biphenyl in the expected planar conformation would be closer together than the sum of their van der Waals radii. Hence the two rings twist out of the planar conformation to enable the two chlorine atoms to move further apart so that the repulsive force between them is reduced (Figure 5.3). Such effects are commonly called steric effects. The interactions that are usually considered in the discussion of steric effects are between atoms that are not bonded to a common atom as in Figure 5.3. It has been less commonly recognized that the repulsive interactions between ligands bonded to a common atom, as in AX_n molecules, can be important in determining the bond lengths and angles in many molecules. Such ligands are referred to as **geminal** ligands from the Latin *gemini* (twins). We discuss the importance of repulsive interactions between germinal ligands in determining molecular geometry in the following section.

◆ 5.2   Ligand-Ligand Interactions

In an early electron diffraction investigation of the structure of 2-methylpropene (isobutylene), Bartell and Bonham (1960) found that the three terminal carbon atoms are arranged in an almost perfect equilateral triangle around the central carbon despite the considerable difference in the single and double bond lengths (Figure 5.4). This result led Bartell to suggest that the terminal carbon atoms are close-packed around the central carbon atom. He then

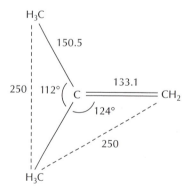

Figure 5.4 Geometrical parameters for 2-methylpropene (isobutene) determined by Bartell in 1960. The three terminal carbon atoms lie at the corners of an equilateral triangle.

suggested that this close packing is responsible for the short length of 150.5 pm of the single C—C bonds in $(CH_3)_2C=CH_2$ compared to the length of 154 pm for a C—C bond in a four-coordinated molecule such as ethane. Three ligands can pack more closely than four, resulting in correspondingly shorter bonds. Bartell pointed out that since this purely steric effect could explain these short bonds, it is not necessary to resort to electronic explanations such as a change in hybridization from sp^3 to sp^2, which in any case is a description rather than an explanation. He followed up this suggestion by showing that he could account for the interligand distances in a variety of substituted ethenes and ketones by attributing a constant nonbonded radius to each of the ligand atoms, assuming that they could be regarded as hard spheres. The radii that he determined are given in Table 5.1 (Bartell 1960). Some examples of the agreement between interligand distances and experimental values for the molecules that he studied are given in Figure 5.5.

Later Glidewell (1975, 1976) extended Bartell's radii to ligands that had not been studied by Bartell. He called them **1, 3 radii** because they are the radii relevant to the first and

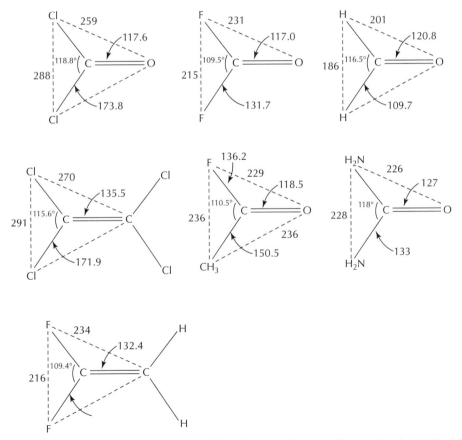

Figure 5.5 Interligand distances (pm) in some of the substituted ethenes and ketones from which Bartell deduced the ligand (1,3) radii in Table 5.1.

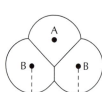

Figure 5.6 The intramolecular 1,3 radius $r_{1,3}$ of B in an angular molecule AB_2.

third atoms in a sequence of three bonded atoms (Figure 5.6). Some of his values are also given in Table 5.1. These 1,3 radii are smaller than the corresponding van der Waals radii. Atoms bonded to a common atom (geminal atoms) are squeezed together more strongly than atoms in separate molecules. The attractive force pulling the atoms toward the central atom and therefore toward each other in a molecule is much stronger than the weak van der Waals forces acting between separate molecules or between atoms in the same molecule that are not bonded to the same atom. Hargittai (1985) has drawn attention to the very nearly constant distances between two given ligands in a number of related molecules, as would be expected from Bartell's close-packing model, but he noted that this constant interligand distance was not always equal to the sum of the Bartell radii. For example, Hargittai found a very nearly constant value of 248 pm for the O···O distance in a wide variety of $XYSO_2$ molecules (Table 5.2), but he noted that this distance is considerably longer than the value of 226 pm obtained from twice the 1,3 radius of oxygen.

Despite the apparent success of Bartell's radii and the appealing simplicity of the model, the importance of ligand–ligand interaction in determining geometry was not widely accepted. An important reason for the lack of enthusiasm for the model was that when it was applied to molecules with central atoms other than the carbon atom, interligand distances often did not agree with the sum of the Bartell radii, as we have just seen in the case of the SO_2 group. This disagreement arises because *the ligand radius of a ligand atom depends on the nature of the atom to which it is bonded,* as we discuss in the following section. The 1,3 radius of oxygen bonded to sulfur, for example, is not the same as the radius deduced by Bartell because the latter applies only to oxygen bonded to carbon.

Table 5.2 SO Bond Lengths, OSO Bond Angles, and O···O Interatomic Distances in Some $XYSO_2$ Molecules

Molecule	S=O Bond Length (pm)	OSO Angle (°)	O···O (pm)
FSO_2F	139.8	125.1	248
FSO_2OCH_3	141.0	124.4	248
FSO_2CH_3	141.1	123.1	248
$ClSO_2Cl$	140.5	123.5	249
$ClSO_2C_6H_5$	141.8	122.3	248
$ClSO_2CH_3$	142.5	120.8	248
$ClSO_2CF_3$	141.6	122.4	248
$ClSO_2CCl_3$	142.1	121.5	248
$CH_3SO_2CH_3$	143.6	119.7	249

◆ 5.3   The Ligand Close-Packing (LCP) Model

In 1997 and 1998 Gillespie and Robinson and their colleagues published the results of an extensive study of interligand distances in a wide variety of AX_n molecules in which A is from the second period and X is fluorine, chlorine, oxygen, or a hydroxy group. They showed that the distance between two given ligands is remarkably constant for a given central atom A. Some examples of F⋯F interligand distances in some molecules of beryllium, boron, and carbon are given in Table 5.3. We see that the F⋯F distance is very constant in these molecules and independent of whether the central atom is three- or four-coordinated and of the presence of other ligands. But the average F⋯F distance decreases as the central atom changes from beryllium to boron to carbon. Similarly, constant Cl⋯Cl distances are found in the chlorides of these elements (Table 5.4) and constant C⋯C distances are found in molecules with BC_n and CC_n groups (Table 5.5).

These constant intramolecular distances between two X ligands are consistent with the model of ligands as hard objects tightly packed around the central atom A. Each ligand can then be considered to be touching its neighbors and can be assigned a nonbonded radius, which is given by half the ligand–ligand distance. This nonbonded radius we call the **intramolecular ligand radius**, or simply the **ligand radius**. On this basis we can assign a

Table 5.3   Bond Lengths, Bond Angles, and F⋯F Distances in Some Molecules Containing BeF_n, BF_n, and CF_n Groups

Molecule	Coord. No.	A—F (pm)	FAF (°)	F⋯F (pm)
BeF_3^-	3	149	120	258
BeF_4^{2-}	4	155.4	109.5	254
				Mean 256
F_3B	3	130.7	120.0	226
F_2B—OH		132.3	118.0	227
F_2B—NH_2		132.5	117.9	227
F_2B—Cl		131.5	118.1	226
F_2B—H		131.1	118.3	225
F_4B^-	4	138.2	109.5	226
F_3B—CH_3^-		142.4	105.4	227
F_3B—CF_3^-		139.1	109.9	228
F_3B—PH_3		137.2	112.1	228
				Mean 226
CF_3^{+a}	3	124.4	120	216
F_2C=CF_2		131.9	112.4	219
F_2C=CCl_2		131.5	112.1	218
F_2C=CH_2		132.4	109.4	216
F_2C=CHF		133.6	109.2	218
F_4C	4	131.9	109.5	215
F_3C—CF_3		132.6	109.8	217
F_3C—OF		131.9	109.4	215
F_3CO^-		139.2	101.3	215
				Mean 216

[a]Ab initio calculated structure.

Table 5.4 Bond Lengths, Bond Angles, and Cl···Cl Distances in Some Molecules Containing $BeCl_n$, BCl_n, or CCl_n Groups

Molecule	Coord. No.	A—Cl (pm)	ClACl (°)	Cl···Cl (pm)
$Cl_2Be(NCMe)_2$	4	197.8	116.8	337
$Cl_2Be(OEt_2)_2$		197.8	116.6	337
			Mean	337
BCl_3	3	174.2	120.0	301
$Cl_2B—BCl_2$		175.0	118.7	301
BCl_4^-	4	183.3	109.5	299
$H_3N—BCl_3$		183.8	111.2	303
$C_5H_5N—BCl_3$		183.7	110.1	301
$Me_3N—BCl_3$		183.1	109.3	299
$Ph_3P—BCl_3$		185.1	109.5	302
			Mean	301
Cl_2CO	3	173.8	111.8	288
$Cl_2C=CH_2$		171.8	112.4	286
CCl_4	4	177.1	109.5	289
H_2CCl_2		176.5	112.0	293
F_2CCl_2		174.4	112.5	290
Me_2CCl_2		179.9	108.3	292
$Cl_3C—CCl_3$		176.9	108.9	288
Cl_3CH		175.8	111.3	290
Cl_3CF		176	109.7	291
			Mean	290

Table 5.5 Bond Lengths, Bond Angles, and C···C Distances in Some Molecules Containing BC_n and CC_n Groups

Molecule	Coord. No.	Bond Length (pm)	Bond Angle (°)	C···C (pm)
$B(CH_3)_3$	3	157.8	120.0	273
$(CH_3)_2BNCO$		156.3	123.6	276
$(C_6H_5)_2BCl$		155.9	123.3	274
$B(CH_2CH_3)_3$		157.3	120.0	273
$B(C_6H_5)_3$		158	120.0	274
			Mean	274
$(H_3C^1)_2C^2=C(CH_3)_2$	3	1: 150.5	1-1: 113.2	251
		2: 133.6	1-2: 123.4	250
$(H_3C^1)_2C^2=CH_2$		1: 150.7	1-1: 115.8	255
		2: 134.2	1-2: 122.1	249
$C_3H_8-C_7H_{16}$	4	153.1 –153.9	111.9–112.9	254–255
$(CH_3)_2CHCl$		152.7	112.7	254
$(HCCl_2)_2CH_2$		152.7	114.2	256
$(H_2CCl)_2CH_2$		153.1	111.6	253
$(BrCH_2)_2CH_2$		152.7	111.4	252
Diamond		154.4	109.5	252
			Mean	254

Table 5.6   Ligand Radii (pm)

Ligand	Be	B	C	N
		Central Atom		
H	—	110	90	82
C	—	137	125	120
N	144	124	119	—
O	133	119	114	—
F	128	113	108	108
Cl	168	151	144	142

value of 108 pm for the radius of an F ligand bonded to carbon, 113 pm for the radius of an F ligand bonded to boron, and a radius of 128 pm for a F ligand bonded to beryllium. Table 5.6 gives values of the ligand radii for H, F, Cl, O, N, and C bonded to Be, B, C and N.

Why does the radius of a given ligand depend on the atom to which it is bonded? With decreasing electronegativity of the central atom, the ligand acquires a greater share of the bonding electrons, thus increasing its negative charge. The size of the atom and therefore its ligand radius increases as it acquires more electron density up to the limit at which it becomes a true ion. It then has its maximum radius, which would be expected to be the same as its ionic crystal radius. We see in Table 5.6 that the ligand radius of a given ligand decreases from left to right across the periodic table as the electronegativity of the central atom A increases and the ligand charge decreases. For example, the ligand radius of fluorine bonded to beryllium, boron, and carbon decreases from 128 to 113 to 108 pm. These radii correlate well with the charges on fluorine determined by the analysis of the electron density distribution (Chapter 6), which are -0.88 in BeF_2, -0.81 in BF_3, and -0.61 in CF_4. As we can see in Table 5.6, the radii for ligands attached to the weakly electronegative Be atom are close to the ionic radii (Table 2.3) consistent with the expected high negative charges on the much more electronegative ligands in these molecules. The charge on a given ligand atom bonded to the same central atom also varies from molecule to molecule. As we shall see in Chapter 8, however, this variation is much smaller than the change in the ligand charge as the central atom changes, so it has only a very small and usually negligible effect on the ligand radius.

The ligand radii in Table 5.6 for ligands bonded to carbon agree well with Bartell's values, as would be expected, but the other radii are different. This is the reason for the lack of success of the 1,3 radii when applied to a range of molecules. Glidewell had assumed that the 1,3 radius was independent of the central atom, so the radii he obtained were not a consistent set, since the radii for different ligands were obtained for different central atoms.

The constancy of the ligand radii, in contrast to van der Waals radii, suggests that geminal ligands on molecules of period 2 elements are squeezed together almost to their limit of compressibility. The repulsive interaction between two atoms is usually represented by a steeply rising potential such as that shown in Figure 5.7. This potential is often approximately represented by a function of the type

$$V = C/r^{12}$$

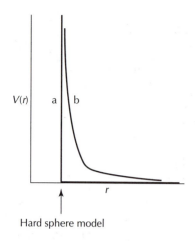

$V(r)$ a   b

r

Hard sphere model

Figure 5.7 The variation of the potential energy as two non-bonded atoms approach each other: curve a, the hard sphere model; curve b, a potential of the form $V = C/r^{12}$.

where V is the potential energy, r the distance between the two atoms, and C a constant. This potential energy curve is very steep at distances corresponding to interligand distances but much less steep at distances corresponding to the distances between atoms in separate molecules, or to atoms in the same molecule that are not bonded to a common atom. Such atoms are therefore much more compressible than geminal atoms. Consequently distances between two atoms in two separate molecules that are "touching" each other in the solid state are much more variable than the distances between the same two atoms when they are geminal atoms in the same molecule.

◆ 5.4   Bond Lengths and Coordination Number

For AX_n molecules with no lone pairs in the valence shell of A, both the VSEPR model and the LCP model predict the same geometries, namely AX_2 linear, AX_3 equilateral triangular, AX_4 tetrahedral, AX_5 trigonal bipyramidal, and AX_6 octahedral. Indeed Bent's tangent sphere model can be used equally as a model of the packing of spherical electron pair domains and as a model of the close packing of spherical ligands around the core of the central atom.

An important consequence of the LCP model is that bond lengths are expected to increase with increasing coordination number from two to three to four to six.

Bond lengths in AX_n molecules increase with increasing coordination number n.

This variation of bond length with coordination number is shown by the data in Tables 5.3 and 5.4 and in Figure 5.8. Because the two ligands in an AX_2 molecule are not close-packed, the bond distance in a two-coordinated molecule is not restricted by interactions between the ligands. Thus we can think of it as the "natural" bond length for these two atoms. In three- and four-coordinated molecules the bonds are longer because ligand–ligand repulsions prevent them from reaching this shorter "natural" bond length. Although in a two-coordinated

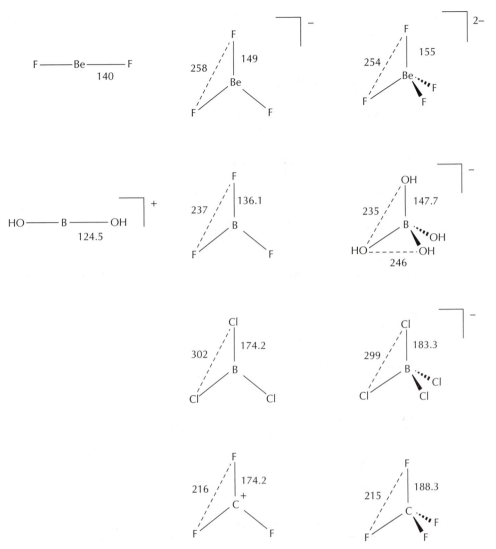

Figure 5.8 Bond lengths increase with increasing coordination number from 2 to 3 to 4.

molecule the ligands are not close-packed (i.e., are not "touching" each other), the linear geometry of the molecule can still be attributed to ligand–ligand interactions, which are min-imized in this geometry.

The increase in bond length with increasing coordination number provides a simple steric explanation of why the bond length in BF_3 is shorter than in BF_4^-. We do not therefore need the back-bonding model (described in Chapter 2) to explain this bond length difference. Ac-cording to this model the bonds in BF_3 are considered to have some double bond character as a consequence of back-bonding, while the bonds in BF_4^- are regarded as single bonds. As we will discuss in Chapter 9, the large differences between the bond lengths of molecules such as SiF_6^{2-} (169.4 pm) and SiF_4 (155.5 pm) and between PCl_6^- (212 pm) and PCl_4^+ (192 pm) can be similarly explained in terms of close packing of the ligands.

◆ 5.5 Molecules with Two or More Different Ligands

Interligand distances between two different ligands in a molecule are given to a very good approximation by the sum of the appropriate radii.

Examples in addition to those in Figure 5.5 are given in Tables 5.7–5.10. In these tables we see that the F⋯Cl, F⋯O, Cl⋯O, C⋯F, and C⋯Cl interligand distances are all very nearly constant, and their mean value in each case agrees well with the sum of the ligand radii. This good agreement provides further strong evidence for the LCP model and confirms the validity and usefulness of the ligand radii. The interligand distances remain constant in these molecules *despite considerable differences in the bond lengths and bond angles.* When a ligand X in an AX_n molecule is replaced by a ligand Y, there will in general be a deviation from a regular geometry, that is, from the ideal bond angles of 90, 109.5, and 120°. There

Table 5.7 Interligand Distances in Some Chlorofluorocarbon Molecules

Molecule	Bond Lengths (pm)		Bond Angle (°)	F⋯Cl (pm)
	C—F	C—Cl		
$FCCl_3$	133	176	109.3	253
F_2CCl_2	134.5	174.4	109.5	253
F_3CCl	132.8	175.1	110.4	254
F_2CHCl	135.0	174.7	110.1	255
$FClCO$	133.4	172.5	108.8	250
			Mean	253
			Sum of ligand radii	252

Table 5.8 Interligand Distances in Some Oxofluorocarbon Molecules

Molecule	Bond Lengths (pm)		Bond Angle (°)	O⋯F (pm)
	C—F	C—O		
$(CF_3)_2O$	132.7	136.9	110.2	221
CF_3O^-	139.2	122.7	116.2	223
CF_3OF	131.9	139.5	109.6	222
F_2CO	131.7	117.0	126.2	222
FCH_3CO	134.8	118.1	121.7	221
$FClCO$	133.4	117.3	123.7	221
$FBrCO$	131.7	117.1	125.7	222
			Mean	222
			Sum of ligand radii	222

Table 5.9 Interligand Distances in Some Fluorocarbon Molecules

Molecule	Bond Lengths (pm)		*Bond angle for CCF*	*C⋯F (pm)*
	C—F	*C—C*		
F₃C—CF₃	132.6	154.5	109.8	234
(CF₃)₃CH	133.6	156.6	110.9	237
(CF₃)CCl	133.3	154.4	111.0	237
H₃CCOF	136.2	150.5	110.5	236
F₂C=CF₂	131.9	131.1	123.8	232
F₂C=CCl₂	131.5	134.5	124.0	235
F₂C=CH₂	131.6	132.4	125.2	234
trans-FCH=CFH	134.4	132.9	119.3	231
			Mean	234
			Sum of ligand radii	232

are two reasons for the effect of the ligand Y on the bond angles: (1) it has a different size (ligand radius), and (2) it forms a bond of a different length. Generally, larger ligands form longer bonds, and so these two factors often approximately cancel and their combined effect on the bond angles may be relatively small, as shown by the nearly tetrahedral bond angles in some chlorofluorocarbon molecules (Table 5.7). The largest effect on the bond angles is seen when a ligand is replaced by another ligand of comparable size that nevertheless forms a shorter bond. This is shown clearly when we compare $F_2A=O$ molecules with the corresponding AF_3 molecules and $F_3A=O$ molecules with the corresponding AF_4 molecule as in Figure 5.9. In F_2CO, despite the similar ligand radii of O and F, the FCO angle is 125.2° and the FCF angle is 107.6°. This difference in bond angles is a consequence of the short length of the CO bond (117.0 pm) compared to the CF bond (132.0 pm) and the requirement that the ligands remain close-packed. The close packing of the ligands is shown by the interligand distances, which are close to the sum of the ligand radii. Similarly, in POF₃ the FPF angles are smaller than the tetrahedral angle owing to the short length of the PO bond. We note also that the AF bond length increases from CF_3^+ to F_2CO and from PF_4^+ to F_3PO. In both cases, for the more strongly bonded oxygen ligand to reach its equilibrium distance

Table 5.10 Interligand Distances in Some Chlorocarbon Molecules

Molecule	Bond Lengths (pm)		CCl (°)	C⋯Cl (pm)
(CH₃)₂CCl₂	152.3	179.9	108.9	271
CH₃CH₂Cl	152.8	174.6	110.7	274
CH₃COCl	150.8	179.8	112.2	275
OClC-COCl	153.6	174.6	111.7	272
Cl₂C=CCl₂	135.5	171.9	122.2	270
H₂C=CHCl	135.5	172.8	121.1	269
Cl₂C=C=CH₂	132.6	173.3	122.2	269
			Mean	271
			Sum of ligand radii	271

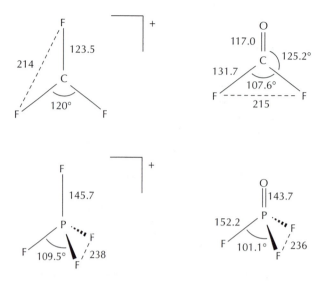

Figure 5.9 Multiple bonds and bond angles. Because multiple bonds are short and strong, they push away the singly bonded ligands, increasing the angle they make with the neighboring single bonds and increasing the length of these bonds. Alternatively, we can imagine that the central atom is pulled toward the more strongly bound doubly bonded oxygen and away from the center of the equilateral triangle of the AX_3 molecule or the tetrahedron of the AX_4 molecule when an X ligand is replaced by an O ligand.

from the central atom, it pushes away the more weakly bound fluorine ligands, increasing the length of the A—X bond and decreasing the FAF bond angle. The short CF bond in CF_3^+, like that in BF_3 discussed in Chapter 2, was formerly attributed to double-bond character arising from back-donation of lone pair electrons from fluorine to the central atom. However, ligand close packing provides an alternative explanation that must be at least an important factor in the bond shortening and may be the only cause. Because multiple bonds are considerably shorter than comparable single bonds, the angles between them are generally larger than between single bonds, as we saw in Figure 4.19. This explanation is an alternative to the larger double bond domains of the VSEPR model. We will discuss further examples of bond angles in molecules containing multiple bonds in Chapters 8 and 9.

◆ 5.6 Bond Angles in Molecules with Lone Pairs

It might at first sight appear that the LCP model is not consistent with the effect of lone pairs on geometry that is so nicely accounted for by the VSEPR model. For example, the packing of three ligands around a central atom might be expected to lead to the same planar triangular geometry for both an AX_3 and an AX_3E molecule. This would indeed be true for a hypothetical purely ionic molecule consisting of three spherical X^- anions packed around a spherical central cation A. However, we will see that if the central atom in a real molecule has unshared electrons, the central atom is not spherical. This is because the bonding electron density and the negative charge on the ligands interact with the unshared valence density, which would be spherical in the free ion, pushing it to one side of the core and creating lone pairs even in a very ionic molecule. This is illustrated in Figure 5.10 for the case of PF_3 in which we imagine three spherical F^- ions approaching a spherical P^{3+} ion that consists of a spherical P^{5+} core and two nonbonding electrons surrounding the core in a spherical distribution. As the F^- ions approach the P^{3+} ion, they distort the spherical distri-

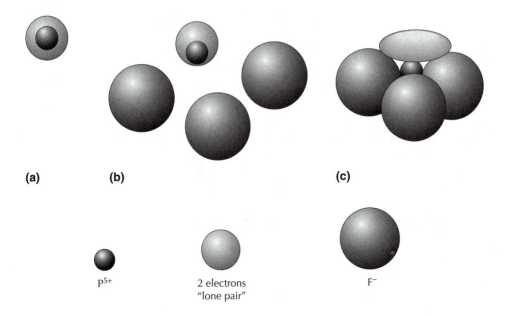

(a)   (b)   (c)

P⁵⁺   2 electrons   F⁻
                   "lone pair"

Figure 5.10 Representation of the formation of the lone pair in the PF_3 molecule. (a) An isolated P^{3+} ion consisting of a P^{5+} core surrounded by two nonbonding electrons in a spherical distribution. (b) Three approaching F^- ions distort the distribution of the two valence shell electrons pushing them to one side of the P^{5+} core. (c) When the F ligands reach their equilibrium positions, the two nonbonding electrons are localized into a lone pair, which acts as a pseudo-ligand giving the PF_3 molecule its pyramidal geometry.

bution of the two nonbonding electrons, pushing them to one side and into a more localized distribution—in other words, causing them to adopt the distribution of a typical nonbonding domain. The three F ligands then avoid the lone pair domain, thus adopting a trigonal pyramidal geometry as predicted by the VSEPR model. The lone pair domains occupy part of the valence shell, preventing the ligands from occupying this space. The lone pairs can therefore be regarded as pseudoligands, and as a result, the predicted AX_2E, AX_3E, and AX_2E_2 geometries are the same as predicted by the VSEPR model.

Because of the tendency of the lone pair density to spread out as much as possible, the ligands in AX_2E, AX_3E, and AX_2E_2 molecules are generally pushed into contact, giving the same ligand–ligand distances as in AX_3 and AX_4 molecules (Figure 5.11). On the basis of the assumption that lone pair domains are larger than bonding pair domains, the VSEPR model predicts that in the presence of a lone pair, the bond angles will be smaller than in the corresponding regular polyhedron. For example, the VSEPR model predicts a bond angle smaller than 109.5° in an AX_3E molecule and smaller than 90° in an AX_5E molecule, but it cannot make more quantitative predictions. The LCP model, however, enables us not only to predict the interligand distances in such molecules but to also quantitatively predict bond angles when the bond lengths are known.

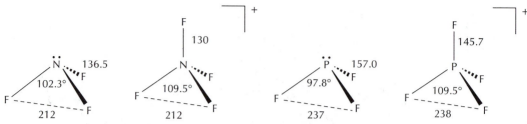

Figure 5.11 Bond angles and intramolecular contact distances in AX$_3$E and AX$_4$ molecules.

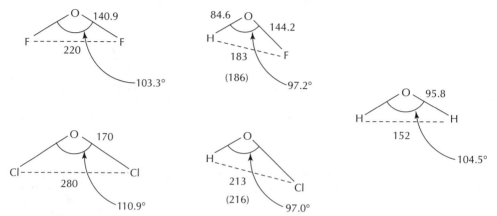

Figure 5.12 Bond angles and intramolecular contact distances in HOX molecules.

For example, the LCP model provides a simple explanation of the initially surprising observation that the bond angles in HOX molecules are smaller than in both the H$_2$O and the X$_2$O molecules (Figure 5.12). The small angle in the XOH molecules is a consequence of the constant interligand distance and the different bond lengths. Knowing the bond lengths, we can predict the bond angle. In contrast, the VSEPR model, on the basis of electronegativity considerations, would predict, incorrectly, that HOX molecules will have bond angles between those of the H$_2$O and X$_2$O molecules.

◆ 5.7 Weakly Electronegative Ligands

If the electronegativity of the ligands X is much less than the electronegativity of the central atom A, the electrons in the valence shell of A are not well localized into pairs and therefore have a small or zero effect on the geometry. In such molecules the bonds are very ionic in the sense A$^-$X$^+$, and the central atom A is essentially an anion with a spherical electron density distribution. In this case the VSEPR model is not valid, and the geometry of the molecule is determined by ligand–ligand repulsions.

For example, the LiOH and Li$_2$O molecules are linear, not angular like H$_2$O, because the Li ligand is not sufficiently electronegative to localize the eight electrons in the valence

Table 5.11 Calculated Geometries of the
$A(OH)_n$ Molecules of Period 2

Molecule	A—O (pm)	OAO (°)	AOH (°)
LiOH	158.2	—	180.0
Be(OH)$_2$	142.3	180.0	134.5
B(OH)$_3$	136.8	120.0	112.8
C(OH)$_4$	139.3	103.6, 112.5	106.9
N(OH)$_3$	141.3	103.8	102.6
O(OH)$_2$	144.4	100.3	98.7
FOH	143.2	—	98.6

shell of oxygen into four tetrahedrally arranged pairs. So in these molecules the ligand–ligand interactions dominate and give the observed linear geometry. Table 5.11 shows that as the electronegativity of A increases in a series of $A(OH)_n$ molecules, the increasing localization of the electrons on oxygen into pairs leads to a decrease in the calculated AOH bond angle from 180° to a value less than the tetrahedral angle in FOH, as expected from the VSEPR model.

In general, when the localization of the electrons on the central atom is so weak that ligand–ligand repulsions dominate the geometry, bond angles in OX_2E_2 and NX_3E molecules are larger than tetrahedral and may reach the maximum possible angles of 180° in OX_2E_2 molecules and 120° in NX_3E molecules (Table 5.12). For example, the Li_2O molecule has a linear geometry and $N(SiH_3)_3$ has a planar geometry . The effect of increasing the electronegativity of X is shown by the bond angles in the molecules Li_2O (180°, calc), $(CH_3)_2O$ (111.7°), F_2O (103.3°), and in the molecules Na_2O (180°, calc.), $(SiH_3)_2O$ (144.1°), Cl_2O (110.9°). As the nonbonding electrons become increasingly localized into lone pairs with increasing electronegativity of the ligands in each of these series of molecules the bond angles approach those predicted by the VSEPR model. The VSEPR description of the molecules AX_2E_2 and AX_3E when they have very weakly electronegative ligands X is not strictly valid because the well-localized lone pairs denoted by E are not present. Molecules such as $N(SiH_3)_3$ and $O(SiH_3)_2$ are often cited as exceptions to the VSEPR model and are used to question the validity of the model. However, they cannot be considered to be true exceptions because the model is not expected to be applicable to these molecules.

Table 5.12 Bond Angles and Bond Lengths in NX_3E and OX_2E_2 Molecules

Molecule	N—X (pm)	XNX (°)	Molecule	O—X (pm)	XOX (°)
NF$_3$	136.5	102.3	F$_2$O	140.0	103.3
NCl$_3$	175	106.8	H$_2$O	95.8	104.3
NH$_3$	101.5	107.2	Cl$_2$O	170.0	110.9
N(CH$_3$)$_3$	145.8	110.9	(CH$_3$)$_2$O	141.0	111.7
N(CF$_3$)$_3$	142.6	117.9	(SiH$_3$)$_2$O	163.4	144.1
N(SiH$_3$)	173.4	120.0	(Me$_3$Si)$_2$O	163	148
			Li$_2$O	160	180

◆ 5.8 Ligand–Ligand Interactions in Molecules of the Elements of Periods 3–6

Almost all the examples discussed in the preceding sections have been drawn from molecules of the period 2 elements because they have been more intensively studied and because these provide good illustrations of the LCP model. The bonds in molecules of the period 3 elements are longer and weaker than the corresponding bonds formed by the period 2 elements. Consequently, in these molecules the interligand distances are generally larger than in the corresponding period 2 molecules because the ligands are not squeezed so tightly together. The potential energy curve in Figure 5.7 is less steep at the interligand distances in these molecules than in the period 2 molecules. The ligands are therefore softer and more compressible than the same ligand in a period 2 molecule, which in turn give them a slightly more variable ligand radius. Consequently interligand distances are somewhat more variable. They have their smallest value in six-coordinated octahedral molecules in which the ligands probably reach the effective limit of their compressibility, at least for molecules of the elements of periods 3 and 4.

Values of the ligand radii for the ligands F, Cl, and O bonded to the period 3 atoms Al, Si, P, and S are given in Table 5.13 for both four- and six-coordinated molecules. We see that the radius for six-coordinated molecules is a little smaller than that for four-coordinated molecules, where the ligands are less tightly packed. Although the concept of a unique ligand radius independent of coordination number is more an approximation for the elements of period 3 and beyond than for period 2 elements, ligand–ligand interactions are still a very important factor in determining the geometry of the molecules of these elements, as we can see for the molecules PF_3, PF_4^+, and POF_3 in Figures 5.9 and 5.11 and as we will see in other examples in Chapter 9.

◆ 5.9 Polyatomic Ligands

In the discussion in the preceding section we assumed that a ligand atom can be assigned a single ligand radius. However, as we saw in Section 5.1, this assumption is not strictly correct even for monatomic ligands, such as F, Cl and $=O$. We saw in Figure 5.2 that the van der Waals radius varies a little with the direction in which it is measured. However, ligand

Table 5.13 Ligand Radii (pm) for F, Cl, and O Ligands Bonded to Central Atoms of Some Period 3 Elements

Ligand	Six-Coordination				Four-Coordination			
	Al	Si	P	S	Al	Si	P	S
F	128	120	111	110	135	127	120	118
Cl	160	151	148	145	172	160	156	155
O	134	132	126	121	140	132	126	124

radii are less variable than van der Waals radii and moreover the direction of intramolecular contacts measured from the bond direction varies only over limited range from approximately 45° in a six coordinated molecule to approximately 30° in a three-coordinated molecule. So we expect the variation of ligand radius with the contact direction for such ligands to be very small and probably less than the accuracy to which ligand radii can be determined. We can therefore attribute a constant unique ligand radius to a monatomic ligand or a linear ligand, such as CN, which have an electron distribution that is symmetrical around the bond axis. However, this is not the case for nonlinear polyatomic ligands, such as the OH ligand, in which the oxygen does not have an electron distribution that is symmetrical around the axis of the bond to the central atom. This causes a variation in the ligand radius with direction which, although small, is significant, and can have some important consequences. In particular the contact distance with another ligand can vary with the orientation of the ligand. This is illustrated by the example of the two possible planar geometries for the $B(OH)_3$ molecule in Figure 5.13. The C_{3h} geometry is the lowest energy geometry and is the experimentally observed geometry. To account for the interligand distances in the two forms of the $B(OH)_3$ molecule we need to assign two radii to the oxygen atom; r_1, in the direction between the bonds to oxygen and r_2, in the opposite direction, that is in the lone pair region. As we can see from Figure 5.13 the three O—-O distances in the C_{3h} geometry are all equal and are equal to the sum of the two radii r_1 and r_2. The three interligand distances in the C_s structure are not all equal, because they are $r_1 + r_1$, $r_2 + r_2$, and $r_1 + r_2$. Consequently the three bond angles are not equal to each other. From the interligand distances in the two structures of $B(OH)_3$ the two radii are found to be 116 pm and 122 pm. Note that the value of 119 pm for the ligand radius of oxygen bonded to boron given in Table 5.5 is an average radius for oxygen in various OX groups and for terminal oxygen atoms as in the BO_3^{3-} ion. It is interesting to note that the radius of oxygen in the direction of the lone pair region is smaller than the radius in the direction between the two bonds. This observation is consistent with the relative shapes of bonding pair domains and lone pair domains. Lone pair domains are more spread out around the inner core than bond pair domains which are more extended towards the ligands. This picture of the electron distribution around an oxygen atom, for example, is confirmed by calculated electron density distributions (Chapter 6).

An important consequence of the nonaxially symmetric electron distribution of the oxygen atom in an angular OX ligand is that $B(OX)_4$ molecules do not have a regular tetrahe-

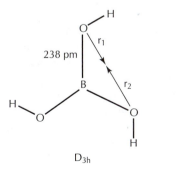

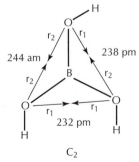

Figure 5.13 The C_{3h} and C_s geometries of the $B(OH)_3$ molecule.

dral geometry. For example, the $B(OH)_4^-$ ion has D_{2d} not T_d symmetry with two OBO angles of 106.2° and four of 111.1°. These deviations from tetrahedral geometry have been shown to be due to the nonaxially symmetrical electron density around the oxygen atom as we discuss in more detail in Chapter 8. Other $A(XY)_4$ and related molecules show similar deviations from tetrahedral geometry because of the nonaxially symmetric electron density around the X atom. The form of the electron distribution around a ligand may be important in understanding other deviations from predicted geometries that have not thus far been easily explained.

◆ 5.10 Comparison of the LCP and VSEPR Models

At first sight the LCP and VSEPR models might appear to be very different and unrelated, but it has perhaps become clear in the preceding discussion that this is not the case. Indeed, the LCP model can be regarded as an extension and a refinement of the VSEPR model in which bond pair–bond pair repulsions are replaced by ligand–ligand repulsions. Both models lead to the same predictions of the general geometry of a molecule. In most cases we cannot distinguish between the effects of bond pair–bond pair repulsions and ligand–ligand repulsions in determining geometry. Indeed the spherical electron pair domain version of the VSEPR model and the ligand close-packing model, in which we consider spherical anion-like ligands to be packed around a spherical cation-like central atom, are mathematically identical and Bent's tangent sphere model applies equally to both. The very important role of lone pairs in determining geometry remains the same in both models. Moreover, the essential role of the Pauli principle in the formation of lone pairs is the same in both models. The importance of the LCP model is twofold: it provides a better understanding of bond angles that is essentially quantitative for molecules of the period 2 elements, and it can predict the geometry of molecules to which the VSEPR model does not apply. This advantage of the LCP model does not, however, mean that electron pair domain interactions are not important. The formation of localized electron pair domains is an expression of the Pauli principle, and these domains are always present in the valence shell of an atom in a molecule in which the attached ligands have a sufficiently great electronegativity. In a molecule in which ligand–ligand interactions were unimportant, the geometrical arrangement of the localized electron pair domains would determine the geometry of the molecule. In a molecule in which there are no well-defined nonbonding electron pairs in the valence shell of the central atom, ligand–ligand interactions dominate the geometry. In the majority of molecules, both electron pair–electron pair and ligand–ligand interactions are important in determining geometry. When the ligands are relatively small compared to the central atom, electron pair–electron pair interactions are probably the more important, but when the ligands are relatively large, no doubt ligand–ligand interactions take precedence. For molecules of the period 2 elements that have small central atoms, ligand–ligand interactions appear to dominate, as we have seen. When it is applicable, the LCP model has the advantage of being able to predict the interligand distances on which the bond lengths and bond angles depend, and thus it gives us a more quantitative understanding than the VSEPR model.

▶ Further Reading

L. S. Bartell, *J. Chem. Educ., 45,* 754, 1968.
   This is a good discussion of the importance of intraatomic interactions in determining molecular geometry.
I. Hargittai, *The Structure of Volatile Sulphur Compounds,* 1985, Reidel, Dordrecht.
   Discusses constant interligand distances and has a large amount of structural information on sulfur containing molecules.

▶ References

L. S. Bartell, *J. Chem. Phys. 32,* 827, 1960.
L. S. Bartell and R. A. Bonham, *J. Chem. Phys. 32,* 624, 1960.
C. Glidewell, *Inorg. Chim. Acta 12,* 219, 1975.
C. Glidewell, *Inorg. Chim. Acta, 20,* 113, 1976.
R. J. Gillespie, I. Bytheway, and E. A. Robinson, *Inorg. Chem. 37,* 2911, 1998.
R. J. Gillespie and E. A. Robinson, *Adv. Mol. Struct. Res. 4,* 1, 1998.
E. A. Robinson, G. L. Heard, and R. J. Gillespie, *J. Mol. Struct. 485–486,* 305, 1999.
E. A. Robinson, S. A. Johnson, T.-H., Tang, and R. J. Gillespie, *Inorg. Chem. 36,* 3022, 1997.
G. L. Heard, R. J. Gillespie, and D. W. H. Rankin, *J. Mol. Struct. 520,* 237, 2000.

CHAPTER 6

THE AIM THEORY AND THE ANALYSIS OF THE ELECTRON DENSITY

■ ■ ■

◆ 6.1 Introduction

We have seen in Chapter 3 that ab initio calculations can give us very accurate solutions of the Schrödinger equation from which we can obtain the molecular wave function, and therefore the electron density distribution, as well as the corresponding energy and equilibrium geometry of a molecule. However, this information, important as it is, does not provide us directly with either the properties of the atoms as they exist in the molecule or of the bonds between them. To obtain this information we have to analyze either the wave function or the electron density. We have discussed some of the difficulties associated with the analysis of the wave function in Chapter 3. The theory of atoms in molecules (AIM), which we describe in this and the following chapter, provides a method for analyzing the electron density distribution (obtained either by an ab initio calculation or by X-ray crystallography) to provide us with information about atoms as they exist in a molecule and on the bonds between them.

The importance of understanding the electron density is made clear by the Hellmann–Feynman theorem. We will see that this theorem shows us that all the properties of a molecule are ultimately determined by the electron density distribution ρ.

◆ 6.2 The Hellmann–Feynman Theorem

The Hellmann–Feynman theorem demonstrates the central role of ρ, the electron density distribution, in understanding forces in molecules and therefore chemical bonding. The main appeal and usefulness of this important theorem is that it shows that the effective force acting on a nucleus in a molecule can be calculated by simple electrostatics once ρ is known. The theorem can be stated as follows:

> The force on a nucleus in a molecule is the sum of the Coulombic forces exerted by the other nuclei and by the electron density distribution ρ.

The only forces operating in a molecule are electrostatic forces. There are no mysterious quantum mechanical forces acting in molecules.

Rather than giving the general expression for the Hellmann–Feynman theorem, we focus on the equation for a general diatomic molecule, because from it we can learn how ρ influences the stability of a bond. We take the internuclear axis as the z axis. By symmetry, the x and y components of the forces on the two nuclei in a diatomic are zero. The force on a nucleus α therefore reduces to the z component only, $F_{z,A}$, which is given by

$$F_{z,A} = Z_A e^2 \int d\tau\, \rho(\mathbf{r}) \frac{\cos\theta_A}{r_A^2} - \frac{Z_A Z_B e^2}{R_{AB}^2} \tag{6.1}$$

where R_{AB} is the distance between the two nuclei, r_A is the distance between nucleus A and a given point with coordinates $\mathbf{r}(x, y, z)$, e is the fundamental charge, Z_A, Z_B are the respective atomic numbers of the nuclei, and θ_A and θ_B are the angles between the position vector $\mathbf{r}$ and the internuclear axis (Figure 6.1). This formula shows that there are two contributions to the force on a nucleus: a term depending on the three-dimensional profile of the electron density throughout the molecule and a nuclear repulsion term. The first term of this expression also embodies the oft-quoted statement, "The electron density acts as glue holding the nuclei together in a stable chemical bond." However, the work of Berlin showed that this is a misleading oversimplification.

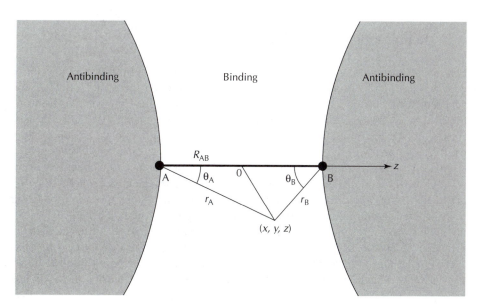

R_{ab} = distance between A and B

Figure 6.1 Binding and antibinding regions for a heteronuclear diatomic molecule consisting of two nuclei A and B with $Z_A = Z_B$. The coordinate system is superimposed. The distance from a point with coordinates (x,y,z) to nucleus A is r_A and to nucleus B is r_B. the distance between the nuclei is R_{AB} To obtain the 3D binding and antibinding regions rotate the figure about the internuclear axis.

Berlin showed that a diatomic molecule can be partitioned into a **binding region** and an **antibinding** region, as shown in Figure 6.1. These regions are separated by two surfaces of revolution given by the function B:

$$B(Z_A, Z_B, \mathbf{r}) = Z_A \frac{\cos \theta_A}{r_A^2} + Z_B \frac{\cos \theta_B}{r_B^2} = 0 \qquad (6.2)$$

All points in space that obey Equation (6.2) together form two surfaces whose shape depends on the values of the atomic numbers of the nuclei in the diatomic moleule. An element of electronic charge situated in either of these two surfaces exerts a zero force on the nuclei. An element of charge at a point in the space where $B(Z_A, Z_B, \mathbf{r}) > 0$ draws the two nuclei together. In other words, charge present in the binding zone, where $B(Z_A, Z_B, \mathbf{r}) > 0$, contributes to the stability of the bound state of a molecule. On the other hand, charge in the antibinding zone, where $B(Z_A, Z_B, \mathbf{r}) < 0$, attracts one nucleus more strongly than the other, and thus the nuclei are drawn apart.

A stable molecule at equilibrium can exist only if the attractive force from the binding region balances both the nuclear repulsive force and the repulsive force due to charge in the antibinding zone. Sufficient electronic charge must therefore be accumulated in the binding region to produce a stable molecule. This requirement combined with the decrease in ρ with increasing distance from a nucleus, leads electronic charge to be most concentrated along the internuclear axis.

◆ 6.3   Representing the Electron Density

We have just seen that the equilibrium geometry and stability of a molecule are determined by the electron density via the forces it exerts on the nuclei. Now we focus on the representation of ρ and on how we can extract chemical information from it. In contrast to the wave function, the electron density is an observable that can be measured by X-ray crystallography (Section 6.5). Moreover, it is easier to understand than the wave function thanks to the Born interpretation discussed in Chapter 3. Nevertheless, ρ has generally been given relatively little attention in books on quantum chemistry in comparison to wavefunctions, energies, and geometries. This neglect is probably due to its superficially trivial appearance: ρ is very high near the nuclei and appears to be featureless elsewhere. Figure 6.2, a relief map of ρ in the molecular plane of SCl_2, is typical of many molecules. The huge peaks near the nuclear positions have been truncated for convenience at 15 au but are in reality about 200 times higher. The density near the nuclei is so large that the outer regions of this density dominate and tend to obscure the relatively very small, but nevertheless important, features of the density in the bonding region.

We can conveniently think of ρ as a gas with a nonuniform density, which is more compressed and therefore more dense in some regions, and less compressed or less dense in other regions. Since the electron density $\rho(x, y, z)$ of a molecule varies in three dimensions, we need a fourth dimension to represent it completely. Nevertheless we can get a good idea of the behavior of ρ by plotting constant electron density envelopes.

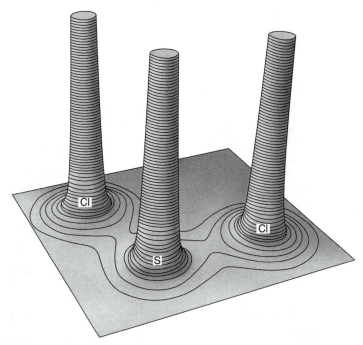

Figure 6.2 Relief map of the electron density in the molecular plane of SCl_2. The vertical direction (z axis) is used to show the value of ρ, which depends on the two coordinates (x,y) describing the molecular plane. The value of ρ at the nuclear positions is of the order of 3×10^3 au but the peaks have been truncated at 15 au. Note the dramatic behavior of the electron density in the vicinity of the nuclei: there are huge peaks appearing on a nearly flat landscape.

Let us consider constant ρ envelopes of SCl_2 for three different values of ρ. The $\rho = 0.001$ au envelope encompasses the whole molecule (Figure 6.3a) and can be regarded as the practical edge of the molecule. In general, this envelope agrees well with the van der Waals surface determined from the molecule's nonbonded interactions with other molecules in the gaseous phase. This surface encloses over 98% of the density in a typical hydrocarbon molecule. As we increase ρ further, the corresponding envelope shrinks and eventually fails to encompass the whole molecule. When $\rho > 0.133$, au the envelope separates into three surfaces, each encompassing only one nucleus (Figure 6.3b). When $\rho = 0.133$ au (Figure 6.3c), these three surfaces touch at only two special points, which we discuss later. Atomic units (au) are described in Box 6.1.

Figure 6.4 shows a third and commonly used way of representing electron density profiles: the two-dimensional contour map. This map for the SCl_2 molecule corresponds to the relief map in Figure 6.2. Although this map is able to show very detailed information, we are restricted to a particular choice of plane, or to a selection of planes. To obtain an approximately equally dense distribution of contour lines, contour values used in this book increase in the nearly geometrical sequence, $10^{-3}, 2 \times 10^{-3}, 4 \times 10^{-3}, 8 \times 10^{-3}, 2 \times 10^{-2}$,

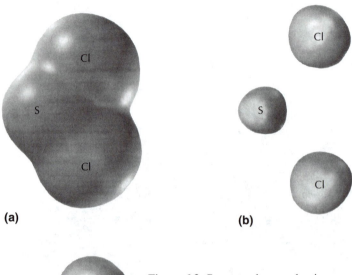

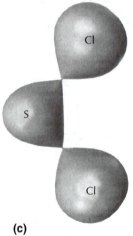

(a)

(b)

(c)

Figure 6.3 Constant electron density envelope maps for SCl_2 for three different contour values (a) $\rho = 0.001$ au, (b) $\rho = 0.200$ au and (c) $\rho = 0.133$ au. (a) This constant density envelope shows the practical outer boundary of the molecule broadly corresponding to the van der Waals envelope. (b) This constant density envelope demonstrates that for higher ρ values the envelope becomes disconnected into three surfaces each encompassing a nucleus. (c) This constant density envlope is plotted at the highest ρ value for which the molecular envelope is still connected or encompasses the whole molecule.

$4 \times 10^{-2}, 8 \times 10^{-2}, \ldots$, au. Each subsequent contour value is then approximately twice the value of the preceding contour value.

In the case of linear molecules, a two-dimensional contour map takes advantage of the cylindrical symmetry of ρ to give an almost complete picture of the electron density. To represent ρ completely, we would need a contour map through the two nuclei and contour maps in an infinite number of planes perpendicular to the molecular axis. Contour maps of ρ for the CO and Cl_2 molecules are given in Figure 6.5. There is a maximum in the electron density at each of the nuclei, and ρ decreases less rapidly along the internuclear axis than in any other direction from either nucleus. In Cl_2 (Figure 6.5a) the electron density reaches a minimum along the internuclear axis at the midpoint, but in CO (Figure 6.5b) the minimum is somewhat closer to the carbon nucleus. This minimum has the same properties as the point between the S and Cl nuclei in SCl_2, where the two constant ρ envelopes have one point in common (see Figure 6.3c, $\rho = 0.133$ au). If we gradually increase the ρ value of an enve-

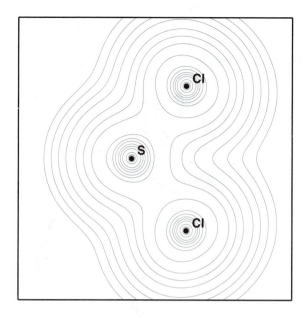

Figure 6.4 Contour plot of the electron density in the molecular plane of SCl_2. The outer contour line corresponds to 0.001 au and the next contour lines correspond to values increasing according to the pattern $2 \times 10^n, 4 \times 10^n, 8 \times 10^n$ where n varies from -3 to 2.

lope map of CO or Cl_2, the envelope first encompasses the whole molecule and then shrinks until it becomes disconnected. The envelope encompasses the two nuclei separately when the ρ contour value is that of ρ at the respective minimum.

What is the behavior of the electron density in a plane perpendicular to the molecular axis? In view of the cylindrical symmetry of a diatomic molecule, we find a pattern of circular contour lines, with values that increase toward the molecular axis. As a result, we can regard the molecular axis as a line along which the electron density is higher than in any direction away from the line. In a two-dimensional contour map or relief map, this line appears as a ridge of higher electron density connecting the peaks that surround the nuclei. In other words, the electron density is more compressed along the molecular axis than along any other line between the nuclei. In Figure 6.6, a relief map of ρ in CO in a plane containing the molecular axis, we see this line of higher density as a ridge connecting the two peaks.

◆ 6.4 Density Difference or Deformation Functions

An early attempt to obtain insight from the molecular electron density was to subtract a reference density from it. The resulting difference density, $\Delta\rho(\mathbf{r})$, introduced by Daudel and others is then simply :

$$\Delta\rho(\mathbf{r}) = \rho_{\text{mol}}(\mathbf{r}) - \rho_{\text{ref}}(\mathbf{r}) \qquad (6.3)$$

where $\rho_{\text{mol}}(\mathbf{r})$ is the electron density of the molecule and $\rho_{\text{ref}}(\mathbf{r})$ is the reference density. A common, reference density distribution is obtained by placing the calculated density of the

▲ BOX 6.1 ▼
Atomic Units

The units we use in daily life, such as kilogram (or pound) and meter (or inch) are tailored to the human scale. In the world of quantum mechanics, however, these units would lead to inconvenient numbers. For example, the mass of the electron is 9.1095×10^{-31} kg and the radius of the first circular orbit of the hydrogen atom in Bohr's theory, the **Bohr radius,** is 5.2918×10^{-11} m. Atomic units, usually abbreviated as au, are introduced to eliminate the need to work with these awkward numbers, which result from the arbitrary units of our macroscopic world. The atomic unit of length is equal to the length of the Bohr radius, that is, 5.2918×10^{-11} m, and is called the **bohr.** Thus 1 bohr $= 5.2918 \times 10^{-11}$ m. The atomic unit of mass is the rest mass of the electron, and the atomic unit of charge is the charge of an electron. Atomic units for these and some other quantities and their values in SI units are summarized in the accompanying table.

There is another important reason for the existence of atomic units, namely, that quantum mechanical expressions, such as the Schrödinger equation, become simpler. When expressed in SI units, the Schrödinger equation for the hydrogen atom is

$$\left(\frac{\hbar^2}{2me} \nabla^2 - \frac{e^2}{4\pi\epsilon_0 r} \right) \psi = E\psi$$

where $\hbar = h/2\pi$, whereas in atomic units it becomes

$$\left(-\frac{1}{2} \nabla^2 - \frac{1}{r} \right) \psi = E\psi$$

Atomic Units

Quantity	Name of Unit	Symbol	Value
Mass	Electron rest mass	m_e	9.1095×10^{-31} kg
Length	Bohr	a_0	5.2918×10^{-11} m
Charge	Elementary charge	e	1.6022×10^{-19} C
Energy	Hartree	E_h	4.3598×10^{18} J
Charge density	au of charge density	$e\, a_0^{-3}$	1.0812×10^{12} C · m^{-3}

neutral spherical ground state atom at each nuclear positions. Although this theoretical reference density is electrostatically binding, it is a hypothetical entity, inasmuch as atoms in molecules normally carry charges and the valence shell electrons are not spherically distributed. Moreover, this hypothetical reference density violates the Pauli exclusion principle because it allows the atomic densities to overlap unchanged. Crystallographers usually refer to this difference density as the **standard deformation density.**

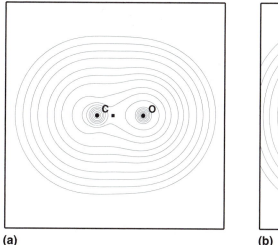

(a)

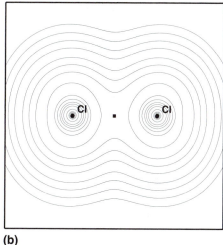

(b)

Figure 6.5 Contour plot of the electron density in a plane containing the nuclei of (a) CO and (b) Cl_2, both drawn at the same scale. Along the internuclear axis, the electron density reaches its minimum value at a point marked by a square. For Cl_2 this is the midpoint.

Figure 6.6 Relief map of the electron density for CO in a plane containing the molecular axis. The electron density falls off more rapidly for displacements perpendicular to the internuclear axis than along the internuclear axis.

Figure 6.7 shows the calculated electron density distributions for the H_2 and N_2 molecules in their equilibrium geometry together with the standard deformation densities. There is clearly a buildup of electron density in the bonding region in both molecules. In the N_2 molecule there is also an increase in the electron density in the lone pair region and a de-

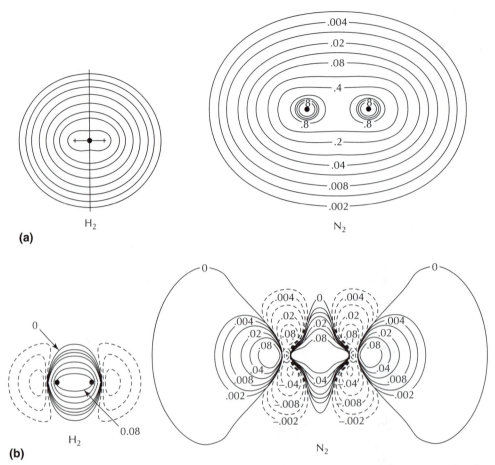

(a)

(b)

Figure 6.7 (a) Contour map of the molecular electron densities H_2 and N_2 in a plane containing the nuclei. (b) Contour maps of a density difference function in the same plane. The solid contour lines indicate positive values, the dashed contour lines negative values. Electronic charge is accumulated in the central bonding region in both molecules and in the lone pair regions of the N_2 molecule. (Reproduced with permission from Bader, Nguyen, and Tal *Rep. Prog. Phys.* **44**, 893, 1981.)

crease in the region between the lone pair density and the bonding density. However, the standard deformation density does not show an accumulation of density in the bonding region for all bonds. Figure 6.8a shows the standard deformation density for the F_2 molecule, which is negative in the bonding region. An F atom with a spherical electron density distribution is obtained by averaging over all the possible configurations $(1s)^2(2s)^2(2p)^5$. This average configuration has $5/3 = 1.66$ electrons per valence p orbital. So when this density is subtracted from the experimental density, 1.66 electrons are subtracted out in the bond region, more than compensating for the accumulation of density due to bond formation and therefore causing a depletion of density in the bond region relative to the spherical atom reference state. The deformation density in the O—O bonding region in hydrogen peroxide and other peroxides is similarly negative. In this case the density of a spherical oxygen atom is

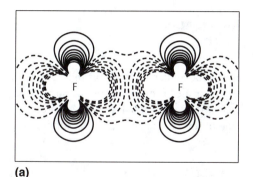 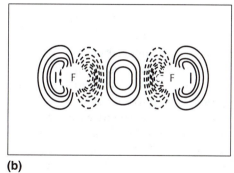

(a) **(b)**

Figure 6.8 Deformation densities for the F_2 molecule. (a) Standard deformation density. Note that there is no charge buildup in the bonding region between the nuclei. (b) Modified deformation density: molecular density minus the density of atoms in the $(1s)^2(2s)^2(2p_x)^2(2p_y)^2(2p_z)^1$ reference state showing a buildup of charge in the bonding region. (Reproduced with permission from P. Coppers [1997].)

the average density of all the configurations $(1s)^2(2s)^2(2p)^4$, and an average of $4/3 = 1.33$ electrons per valence orbital is subtracted out in the bond region, again more than compensating for the accumulation of density due to bond formation.

To remedy the foregoing defect, alternative reference densities have been designed. A more appropriate reference state for fluorine is the valence configuration $(1s)^2(2s)^2 (2p_x)^2(2p_y)^2(2p_z)^1$ with the z axis along the bond direction. In this case no density other than that due to the electrons involved in the bonding is subtracted from the molecular density. This reference state gives the deformation density for the F_2 molecule shown in Figure 6.8b, where we see a buildup of density in the bonding region. This is still only an approximate model for the reference density and so still more refined reference densities have been developed and used, but we do not need to discuss them here. Further details can be found in the book by Coppens (1997).

Clearly the form of a deformation density depends crucially on the definition of the reference state used in its calculation. A deformation density is therefore meaningful only in terms of its reference state, which must be taken into account in its interpretation. As we will see shortly, the theory of AIM provides information on bonding directly from the total molecular electron density, thereby avoiding a reference density and its associated problems. But first we discuss experimentally obtained electron densities.

◆ 6.5 The Electron Density from Experiment

Unlike the wave function, the electron density can be experimentally determined via X-ray diffraction because X-rays are scattered by electrons. A diffraction experiment yields an angular pattern of scattered X-ray beam intensities from which structure factors can be obtained after careful data processing. The structure factors $F(\mathbf{H})$, where $\mathbf{H}$ are indices denoting a particular scattering direction, are the Fourier transform of the unit cell electron density. Therefore we can obtain $\rho(\mathbf{r})$ experimentally via:

$$\rho(\mathbf{r}) = \frac{1}{V} \sum_{\mathbf{H}} F(\mathbf{H}) \exp(-2\pi i \mathbf{H} \cdot \mathbf{r}) \tag{6.4}$$

where V is the unit cell volume.

If we are interested only in the determination of a molecular structure, as most chemists have been, it suffices to approximate the true molecular electron density by the sum of the spherically averaged densities of the atoms, as discussed in Section 6.4. A least-squares procedure fits the model reference density $\rho_{ref}(\mathbf{r})$ to the observed density $\rho_{obs}(\mathbf{r})$ by minimizing the residual density $\Delta\rho(\mathbf{r})$, defined as follows:

$$\Delta\rho(\mathbf{r}) = \rho_{obs}(\mathbf{r}) - \rho_{ref}(\mathbf{r}) \tag{6.5}$$

The model reference density ρ_{ref} is a good approximation to the dominant part of ρ appearing very close to the nuclei, and so $\Delta\rho(\mathbf{r})$ will be very small everywhere and is assumed to be experimental noise. If the peaks in ρ are located, then the nuclear positions are known and the structure is resolved. Because they have no core, hydrogen atoms produce only very small maxima, and thus their positions are difficult to locate with any accuracy. If it is important to locate their positions accurately, this can be done by neutron diffraction. Neutrons are scattered by nuclei rather than electrons, and so the positions of the nuclei are obtained directly. Neutron diffraction is particularly important for the accurate determination of the positions of hydrogen atoms.

However, since the early 1970s crystallographers have gone beyond routine structure determination and have attempted to deduce chemical features such as bonds and lone pairs from the experimentally determined density. Recent advances in experimental techniques, such as dedicated and improved synchrotron X-ray beams, sophisticated photographic detectors, and methods for measuring X-ray diffraction at very low temperatures, have made accurate high-resolution determinations of ρ faster and more reliable than in the past. These methods can give an experimental density of such high quality that $\Delta\rho(\mathbf{r})$ is not just noise but contains chemically relevant information, provided a good reference density is used.

Figure 6.9 shows a contour plot of the deformation density obtained from an experimentally determined density for tetrafluoroterephthalonitrile in the plane of the ring. The density accumulated in the binding regions of all bonds, including the bonds to fluorine, can be clearly seen, as well as the charge buildup in the lone pair zone of nitrogen and fluorine. The buildup of the electron density in the C—F bonding region is rather small. Presumably this small deformation density in the CF bonding region reflects the strong ionic character of the bond as well as the inadequacy of the reference density.

In the next section we will see how the theory of AIM enables us to obtain chemical insight directly from the experimental density as determined by experiment or by calculation, thereby avoiding the need for deformation densities.

◆ 6.6 The Topology of the Electron Density

An important part of AIM is the analysis of the electron density using the branch of mathematics called topology. Topology is the study of geometrical properties and spatial relations

(a)

(b)

Figure 6.9 (a) Standard deformation density of tetrafluoroterephthalonitrile in the molecular plane. Contour interval is 0.1 e Å^{-3}, terminated at 1.5 e Å^{-3}. (b) Molecular diagram with a box around the fragment shown in the deformation map (a). (Reproduced with permission from F. L. Hirshfeld, *Acta Crystallogr., B40*, 613, 1984.)

unaffected by continuous change of shape or size of objects. We can conveniently describe the topology of ρ in terms of its gradient vector field and its critical points.

6.6.1 Gradient Vector Fields

To describe how a quantity varies over a region of space, we use the concept of a *field.* Gravitational, electrical, and magnetic fields are examples of *vector fields,* since a direction is associated with the quantity at each point in space. On the other hand, a *scalar field* is one in which a scalar quantity is used to describe each point in space. For example, the variation in temperature over a given region can be described by a scalar temperature field. The *gradient* of a scalar quantity is a vector that points in the direction in which the scalar quantity is increasing most rapidly. If we were in a submarine at the bottom of the ocean in pitch blackness, looking for a hot spot where lava is escaping from the ocean floor, a thermometer indicating the ocean temperature would allow us to move in the direction of the greatest temperature increase, that is, in the direction of the gradient vector. An important property of a gradient vector is that it is everywhere perpendicular to an envelope of constant scalar value, which in this case would be an envelope of constant temperature, or an isothermal envelope.

There are many other examples of gradient vectors. Perhaps the most familiar example is the direction of steepest ascent on a mountain. The shortest path to the top can be found by taking the direction of steepest ascent at each point. In this way we trace out a path, called a **gradient path,** that is everywhere perpendicular to the contours of constant height. Figure 6.10 shows two of an infinite number of gradient paths for an idealized mountain. The collection of gradient paths is called the **gradient vector field.** Another example is the wind: the gradient of the air pressure coincides with the local wind direction (ignoring the Coriolis effect). The wind blows along a gradient path in a direction perpendicular to the envelopes of constant atmospheric pressure or to the contours called isobars on a two-dimensional map. In the example of the submarine at the bottom of the ocean, the navigator could find the hot spot by following a temperature gradient path that is perpendicular to the isotherms.

The electron density ρ is a scalar quantity that varies through space. The gradient of ρ, denoted by $\nabla\rho$, is a vector that points in the direction of steepest ascent. The operator ∇ is the partial differential operator $\partial\mathbf{u}_x/\partial x + \partial\mathbf{u}_y/\partial y + \partial\mathbf{u}_z/\partial x$, where $\mathbf{u}_x$, $\mathbf{u}_y$, $\mathbf{u}_z$ are unit vectors. If we evaluate $\nabla\rho$ at a given point and follow the vector over an infinitesimally short distance, we will move to a higher ρ value. The path traced by successively following and reevaluating the gradient is a gradient path or line of steepest ascent. The collection of gradient paths constitutes the gradient vector field of the electron density.

What does a typical gradient vector field in the electron density look like? For a free atom, it looks just like the gradient vector field for the idealized mountain illustrated in Figure 6.10. The gradient vector field of ethene in the molecular plane is shown in Figure 6.11b, together with a contour plot of ρ in the same plane (Figure 6.11a). All the gradient paths shown originate at infinity and terminate at a nucleus. The gradient paths do not meet except at a nucleus, and they are always perpendicular (orthogonal) to the contours. Each nucleus acts as an **attractor** for a multitude of gradient paths, which constitute what we call the **basin** of the attractor. The basins do not overlap, and the gradient vector field makes the molecule naturally fall apart into disjoint atomic regions. In other words, the gradient vector field naturally partitions the electron density of a molecule into regions, which as we will see, define the atoms as they exist in the molecule.

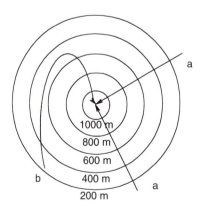

Figure 6.10 The topological map of an idealized mountain represented by the circular contours of constant height on a topological map. Two gradient paths or lines of steepest ascent (a) are shown, together with a path (b) that is not a line of steepest ascent but is an easier route up the mountain. The lines of steepest ascent—gradient paths—cross the contours at right angles.

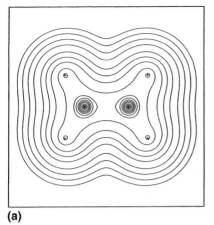 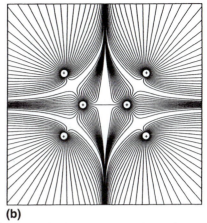

(a)   **(b)**

Figure 6.11 (a) Contour plot of ρ for the molecular plane of the ethene molecule. (b) The gradient vector field of the electron density for the same plane. All the gradient paths shown originate at infinity and terminate at one of the six nuclei.

6.6.2 Critical Points

Now we look at the electron density in another way, namely, in terms of its *extrema*, (i.e. minima, maxima and saddle points), which are called critical points. A **critical point** is an extremum in a function. In a one-dimensional function described by $f(x)$, there are only two types of critical point, maxima and minima, at which the first derivative of $f(x)$, $df(x)/dx$, vanishes (Figure 6.12). Thus at a one-dimensional critical point the slope of $f(x)$ is zero. The two types of critical point can be characterized by the curvature or second derivative of $f(x)$, $d^2f(x)/dx^2$, at these points. At a minimum, the slope changes from negative to positive, and the curvature is positive. Conversely, the slope at a maximum changes from positive to negative, and the curvature is negative. The more pronounced this change of slope for a change in x, the higher the absolute value of the curvature.

A two-dimensional function has extrema, of three types, namely maxima, minima, and saddle points, which are illustrated in Figure 6.12. A three-dimensional function such as ρ has four types of extrema: maxima, minima, and two types of saddle point (see Box 6.2). The maxima in ρ almost always coincide with the nuclear positions, as they do in all the molecules we discuss in this book. We encountered two examples of a saddle point in Section 6.3. Along the internuclear axis in CO and Cl_2 (Figure 6.5) the electron density has a minimum value at a point on the molecular axis between the two nuclei. In a direction perpendicular to the molecular axis, the density at this critical point is a maximum. This point is a saddle point in ρ and is called a **bond critical point.** We showed before that there is a value of ρ such that the corresponding constant density envelope just separates into two envelopes, each encompassing just one nucleus and just touching each other at one point (Figure 6.3c). We now identify this point as the bond critical point. In Figure 6.13, a relief map of the electron density in the molecular plane of methanal, there are three bond critical points, one between the carbon and oxygen nuclei and one between the carbon nucleus and each hydrogen nucleus.

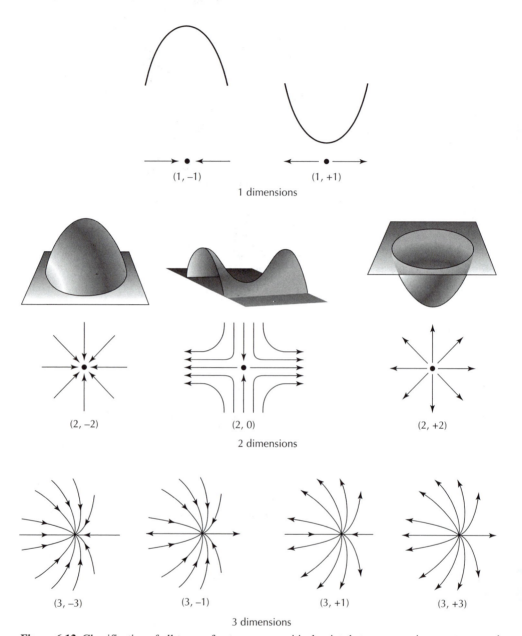

Figure 6.12 Classification of all types of extremum or critical point that can occur in one-, two-, and three-dimensional functions: a one-dimensional function can possess only a maximum or a minimum; a two-dimensional function has maxima, minima, and one type of saddle point; a three-dimensional function may have maxima, minima, and two types of saddle point. The arrows schematically represent gradient paths and their direction. At a maximum all gradient paths are directed toward the maximum, whereas at a minimum all gradient paths are directed away from the minimum. At a saddle point a subset of the gradient paths are directed toward the saddle point, whereas another subset are directed away from the saddle point (see Box 6.2 for more details).

▲ BOX 6.2 ▼
Classification of Critical Points

The classification of critical points in one dimension is based on the curvature or second derivative of the function evaluated at the critical point. The concept of local curvature can be extended to more than one dimension by considering *partial* second derivatives, $\partial^2 f / \partial q_i \partial q_j$, where q_i and q_j are x or y in two dimensions, or x, y, or z in three dimensions. These "partial" curvatures are dependent on the choice of the local axis system. There is a mathematical procedure called matrix diagonalization that enables us to extract local *intrinsic* curvatures independent of the axis system (Popelier 1999). These local intrinsic curvatures are called **eigenvalues.** In three dimensions we have three eigenvalues, conventionally ranked as $\lambda_1 < \lambda_2 < \lambda_3$. Each eigenvalue corresponds to an **eigenvector,** which yields the direction in which the curvature is measured.

The number of nonzero eigenvalues of a critical point is called the **rank *r*.** For example, a maximum in a three-dimensional function has rank 3. The sum of the signs of the eigenvalues is called the **signature *s*.** A value of $(+1)$ is assigned to a positive eigenvalue, or maximum in the corresponding eigenvector direction, and (-1) to a negative eigenvalue, or minimum in the corresponding eigenvector direction. We denote a critical point by (r, s), so a maximum in a three-dimensional function that has a rank of 3 and a signature of -3 is denoted as a $(3, -3)$ critical point. A minimum in two-dimensional function that has a rank of 2 has a signature of $+2$ and is denoted as a $(2, -2)$ critical point. A saddle point of rank 3 that is a maximum in two dimensions and a minimum in one has a signature of -1. So it is a $(3, -1)$ critical point. This explains the notation used in Figure 6.12. In this book we are concerned primarily with critical points of rank 3 and particularly with $(3, -3)$ critical points or maxima and $(3, -1)$ critical points or saddle points. We find $(3, +1)$ critical points only in the center of cyclic molecules such as cyclopropane and $(3, +3)$ critical points or minima only in the center of cage molecules such as P_4. The various types of critical points of rank 3 are summarized in the Table.

We can measure the extent electronic charge is preferentially accumulated by a quantity called the **ellipticity ϵ.** At the bond critical point it is defined in terms of the negative eigenvalues (or curvatures), λ_1 and λ_2 as $\epsilon = (\lambda_1/\lambda_2) - 1$. As $\lambda_1 < \lambda_2 < 0$, we have that $\lambda_1/\lambda_2 > 1$, and therefore the ellipticity is always positive. If $\epsilon = 0$ then we have a circularly symmetric electron density, which is typically found at bond critical points in linear molecules.

Table Box 6.2

	λ_1	λ_2	λ_3	(r, s)
Maximum	—	—	—	$(3, -3)$
Nuclear attractor				
Saddle point	—	—	+	$(3, -1)$
Bond critical point				
Saddle point	—	+	+	$(3, +1)$
Ring critical point				
Minimum	+	+	+	$(3, +3)$
Cage critical point				

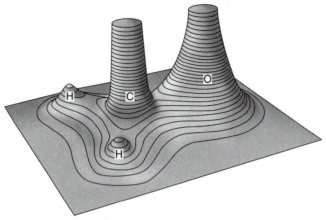

Figure 6.13 Relief map of the electron density for methanal (formaldehyde) in the molecular plane. There is a bond critical point between the carbon and the oxygen nuclei, as well as between the carbon nucleus and each hydrogen nucleus. No gradient path or bond critical point can be seen between the two hydrogen nuclei because there is no point at which the gradient of the electron density vanishes. There is no bond between the hydrogen atoms consistent with the conventional picture of the bonding in this molecule.

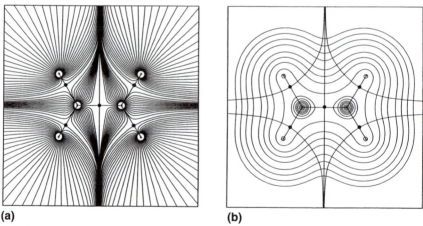

(a)   **(b)**

Figure 6.14 (a) The gradient vector field for ethene, showing a special set of extra gradient paths (cf. Figure 6.11b) that are not attracted to any nucleus. The bond critical points are denoted by a dot. (b) The two sets of special gradient paths superimposed on a contour map of ρ. One set, called bond paths, link the nuclei and the other set mark the boundaries of the atomic basins in this particular plane The black dots denote the bond critical points. These gradient paths, like all gradient paths, are perpendicular to the contour lines.

In Figure 6.14a the gradient vector field of ethene in the molecular plane (shown in Figure 6.11) is augmented by a set of gradient paths that are not attracted to any nucleus. There are two types of paths (Figure 6.14b). Those that start at a bond critical point and terminate at one of the neighboring nuclei and those that start at infinity and terminate at a bond critical point. A set of gradient paths of this latter type define the surface between two atoms that is called the **interatomic surface**. An interatomic surface consists of a bundle of gradient paths originating at infinity and terminating at a bond critical point. As a result, the gradient paths constituting the interatomic surface do not belong to either atom. They lie between two neighboring atoms, hence the name of this surface. Figure 6.14 shows the intersection of five interatomic surfaces (four between C and H, and one between the two C atoms) with the molecular plane in the ethene molecule.

The preceding analysis of the topology of the electron density in a molecule enables us to to define both the atoms and the bonds in a molecule.

6.6.3 Atoms

At the heart of the AIM theory is the definition of an atom as it exists in a molecule. An **atom** is defined as the union of a nucleus and the atomic basin that the nucleus dominates as an attractor of gradient paths. An atom in a molecule is thus a portion of space bounded by its interatomic surfaces but extending to infinity on its open side. As we have seen, it is convenient to take the 0.001 au envelope of constant density as a practical representation of the surface of the atom on its open or nonbonded side because this surface corresponds approximately to the surface defined by the van der Waals radius of a gas phase molecule. Figure 6.15 shows the sulfur atom in SCl_2. This atom is bounded by two interatomic surfaces (IAS) and the $\rho = 0.001$ au envelope. It is clear that atoms in molecules are not spherical. The well-known space-filling models are an approximation to the shape of an atom as defined by AIM. Unlike the space-filling models, however, the interatomic surfaces are generally not flat and the outer surface is not necessarily a part of a spherical surface.

6.6.4 Bonds

Now we focus on the gradient paths, which do not terminate at a nucleus, but rather link two nuclei. For example, the bond critical point between C and H in Figure 6.14 is the origin of two gradient paths. One gradient path terminates at the hydrogen nucleus, the other at the carbon nucleus. This pair of gradient paths is called an **atomic interaction line.** It is found

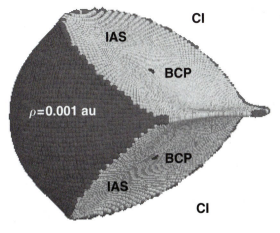

Figure 6.15 Three-dimensional representation of the sulfur atom in SCl_2. This atom is bounded by two interatomic surfaces (IAS) and one surface of constant electron density ($\rho = 0.001$ au). Topologically, an atom extends to infinity on its nonbonded side, but for practical reasons it is capped. Each interatomic surface contains a bond critical point (BCP).

between every pair of nuclei that share a common interatomic surface. When the molecule is in an equilibrium geometry—that is, when the forces on the nuclei vanish—the atomic interaction line is called a **bond path.**

The network of bond paths for a molecule is called its **molecular graph.** It is identical with the network of lines generated by linking together all pairs of atoms that are believed to be bonded to one another according to conventional bonding ideas such as Lewis structures. A bond path can therefore be taken as the AIM definition of a bond.

However, a bond path is not identical to a bond in the sense used by Lewis, that is, to a shared pair of electrons, also usually represented by a line—a bond line. Thus, a molecular graph is not identical to a structural formula or a Lewis structure: For example, double and triple bonds are represented by only one bond path. Figure 6.16 illustrates the molecular graphs of a variety of molecules. There is no bond path between atoms that are not bonded together: for example, no ridge of electron density can be seen between the two hydrogen atoms in ethene or between the two hydrogen atoms in methanal (see Figures. 6.11 and 6.13).

Bond paths are usually but not always straight lines. For example, in a hydrocarbon containing a small ring (e.g., cyclopropane), the bond paths are curved outward from the inter-

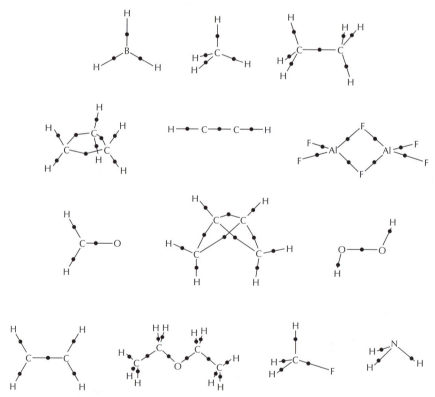

Figure 6.16 Molecular graphs for some molecules in their equilibrium geometries. A bond critical point is denoted by a black dot. The molecule HCCH is ethyne , H_2CO is methanal, and H_2CCH_2 is ethene. [Adapted with permission from Bader [1990], Fig. 2.8.]

nuclear axis, consistent with the bent-bond model of such molecules. A bond path is observed for predominately ionic bonds, for predominately covalent bonds, as well as for weak bonds such as hydrogen bonds and for very weak donor–acceptor bonds such as that in the molecule $BF_3 \cdot CO$, which has a very long and weak B—C bond as we discuss in Chapter 8. The existence of a bond path between two atoms tells us that these atoms are bonded together, but it does not tell us anything about the nature of the bond. We will see later what information about the nature of a bond we can obtain from the critical point on the bond path.

◆ 6.7 Atomic Properties

6.7.1 Determination of Atomic Properties by Integration

Having defined an atom in a molecule, we can, at least in principle, determine any of the properties of an atom in a molecule. The simplest to illustrate is the **atomic volume,** which is simply the sum of all the volume elements that occupy all the space defined by the interatomic surfaces and the $\rho = 0.001$ au contour. More exactly, it is the integral of all the volume elements $d\tau$ over the atomic basin. If we denote the atomic basin by Ω, then the volume of the atom is given by.

$$v(\Omega) = \int_{\Omega} d\tau \qquad (6.6)$$

Because of the irregular shape of an atom in a molecule, this integration is not trivial and can be time-consuming. For many molecules, however, it can now be carried out on a personal computer in a reasonably short time. A discussion of integration procedures is given by Popelier (1999).

Another property that is in principle easily evaluated is the electron population of the atom $N(\Omega)$. This is obtained by integrating the density of a volume element over the atomic basin:

$$N(\Omega) = \int_{\Omega} \rho d\tau \qquad (6.7)$$

The **atomic charge** $q(\Omega)$ is then simply obtained as $Z_{\Omega} - N(\Omega)$, or the electron population subtracted from the charge of the nucleus inside the atomic basin.

The volume, electron population, and charge of the whole molecule can also be obtained by the same method. Table 6.1 gives the atomic volumes, populations, and charges of the atoms in methanal. We see that the electron populations of the atoms and the corresponding charges are additive to four decimal places.

The determination of atomic charges has been a controversial subject because there has been disagreement on how the atoms in a molecule should be defined or, in other words, how the electronic charge should be apportioned between the atoms. Proposed orbital methods for determining atomic charges include the natural bond orbital (NBO) method (Reed et al., 1985, 1988) and the Mulliken method (1985). However, these orbital methods have some unsatisfactory features. There is a certain arbitrary character associated with them because, as we saw in Chapter 3, the choice of orbitals to describe a given molecule is not unique and

Table 6.1 The Volumes, Populations, and Charges of the Atoms in Methanol

Atom		$v(\Omega)$ au	$N(\Omega)$	q
C		66.39	5.019	+0.891
O		138.36	9.060	−1.061
H		50.48	0.960	+0.040
	Total	305.71	16.000	0.000

an atom cannot be completely satisfactorily defined in terms of orbitals which, even when they are localized as much as possible, as in the NBO method, are not totally confined to a single atom in a molecule. There are significant and sometimes large differences between the charges calculated by different orbital-based methods and also between these charges and AIM charges. A comparison of different methods of calculating charges has been given by Wiberg and Rablen (1993). We quote only AIM charges in this book. We will see in Section 6.10 that these charges provide a sound and consistent basis for understanding other properties of the diatomic hydrides, such as the well-established predominately ionic character of LiH and the essentially nonpolar character of the CH bond.

AIM atomic charges are used and discussed extensively in Chapters 8 and 9. Concepts such as covalent and ionic character and the associated concept of electronegativity arose, as we saw in Chapter 1, from descriptions of bonding based on Lewis structures, and from the recognition that since the majority of bonds are polar, the atoms have charges. In principle, the need for these concepts disappears when we have a complete knowledge of the electron density distribution in a molecule, including the atomic charges. But the terms "ionic" and "covalent" have been in use for so long and are so familiar that it is still useful, indeed almost essential, to make use of them in the discussion and interpretation of ρ. We further discuss the meaning and usefulness of these terms in Chapters 8 and 9.

6.7.2 Atomic Dipole Moments

It has often been assumed that atomic charges can be calculated from the measured dipole moment of a diatomic molecule and the bond length. For this assumption to hold, however, the center of negative charge of an atom would have to be situated at the nucleus, in other words, atoms would have to be spherical. But we have seen that atoms in molecules are not spherical, and so the center of negative charge is not centered at the nucleus. Each atom therefore has a dipole moment called the **atomic dipole moment** (Chapter 2).

The atomic dipole moment can be obtained by integrating the moment of a volume element $\rho r_\Omega d\tau$ over the atomic basin. The atomic dipole moment $\mathbf{M}(\Omega)$, where $\mathbf{r}_\Omega$ is a vector centered on the nucleus of the atom, is then

$$\mathbf{M}(\Omega) = \int_\Omega \rho \mathbf{r}_\Omega d\tau \tag{6.8}$$

This moment measures the extent and direction of the shift of an atom's electronic charge cloud with respect to the nucleus. The quantity $\mathbf{M}(\Omega)$ can effectively be regarded as an **intra-atomic dipole moment**. The intra-atomic dipole moment of each atom contributes to the total molecular dipole moment. But there is another contribution, namely, the interatomic dipole arising from point atomic charges located at the nuclear positions, which is called the **charge transfer moment** $\mathbf{M}_{CT}$. The total molecular dipole moment $\mathbf{M}_{mol}$ is given by

$$\mathbf{M}_{mol} = \sum_{\Omega} q(\Omega)\mathbf{X}_{\Omega} + \sum_{\Omega} \mathbf{M}(\Omega) = \mathbf{M}_{CT} + \mathbf{M}_{atom} \tag{6.9}$$

where $\mathbf{M}_{CT}$ is the charge transfer moment, $\mathbf{M}_{atom}$ is the collection of intra-atomic dipoles (due to shifts in charge clouds), and $\mathbf{X}_{\Omega}$ are the nuclear positions of the atoms, measured from a common origin.

As mentioned above and discussed in Chapter 2, atomic charges were often obtained in the past from dipole moments of diatomic molecules, assuming that the measured dipole moment equal to the bond length times the atomic charge. This method assumes that the molecular electron density is composed of spherically symmetric electron density distributions, each centered on its own nucleus. That is, the dipole moment is assumed to be due only to the charge transfer moment $\mathbf{M}_{CT}$, and the atomic dipoles $\mathbf{M}_{atom}$ are ignored.

A good example of the importance of the contribution of the atomic dipoles is provided by the CO molecule, which despite the considerable electronegativity difference between carbon and oxygen has an almost zero dipole moment. The calculated atomic charges are $q(C) = +1.147$ and $q(O) = -1.147$, which are consistent with their electronegativities. The resulting dipole term $\mathbf{M}_{CT} = qr = 2.440$ au is compensated by two large atomic dipoles, counter to the direction of charge transfer, expressed by $\mathbf{M}_{atom}$. The atomic dipole moment of carbon is 1.64 au and that of oxygen is 0.84 au, so that the magnitude of the overall dipole moment $\mathbf{M}_{mol} = 0.040$ au. The large atomic dipole moment on carbon can be associated primarily with the localized and strongly directed lone pair. The asymmetry of the electron density of the carbon atom in CO can be clearly seen in Figure 6.5a.

Another example of the importance of atomic dipoles appeared in Chapter 2, where we attributed the small dipole moment of NF_3 to the moment produced by the lone pair on nitrogen, which makes an important contribution to the atomic dipole on nitrogen and opposes the charge transfer moment due to the electronegativity difference between nitrogen and fluorine.

6.7.3 Additivity of Atomic Properties

An important advantage of the finite atoms defined by AIM is that they do not overlap, which is not generally true for orbital-defined atoms. Each atom has a sharp and well-defined boundary inside the molecule, given by its interatomic surfaces. The atoms fit exactly into each other, leaving no gaps. In other words, the shape and the volume of the atoms are additive. This is true also for other physical properties of an atom, such as the electron population and the charge, as seen in Table 6.2 and as indeed has been shown to be true for all other properties. (Bader 1990, Popelier 1999).

◆ 6.8 Bond Properties

We have discussed the properties of atoms in molecules. What can we find out about the bonds in a molecule? We have seen that the bond path shows us where bonds are located in a molecule, that is, which atoms are bonded together because of the accumulation of elec-

Table 6.2 Atomic and Bond Properties of the Diatomic Hydrides of Periods 2 and 3[a]

Molecule	$v(A)$	$v(H)$	R_e	$r_b(H)$	ρ_b	$q(A)$	$M(A)$	$M(H)$	M_{CT}	M_{mol}
HH	60.5	60.5	1.3792	0.6896	0.2700	0.0000	0.1140	−0.1140	0.0000	0.0000
HLi	36.1	196.7	3.0908	1.7234	0.0379	+0.8869	0.0092	0.4765	−2.7413	−2.2556
HBe	162.9	137.3	2.5469	1.4545	0.0952	+0.8323	1.4387	0.5497	−2.2198	−0.1314
HB	173.3	92.1	2.3163	1.2790	0.1916	+0.6679	1.8201	0.3630	−1.5471	+0.6360
HC	176.4	53.0	2.0941	0.7112	0.2807	−0.0235	0.7327	−0.1685	0.4921	+1.0563
HN	160.7	37.1	1.9347	0.5315	0.3360	−0.3036	0.2456	−0.1943	0.5874	+0.6387
HO	149.5	24.4	1.8111	0.3802	0.3717	−0.5427	−0.1192	−0.1684	0.9829	+0.6953
HF	135.0	15.7	1.7211	0.2848	0.3801	−0.7073	−0.3373	−0.1261	1.2174	+0.7540
HNa	109.0	177.4	3.6176	1.7091	0.0321	+0.7084	0.0568	0.1234	−2.5627	−2.3825
HMg	261.6	145.7	3.3043	1.6195	0.0500	+0.6870	1.4812	0.2424	−2.2701	−0.5465
HAl	267.9	126.4	3.1222	1.5896	0.0758	+0.7677	2.0537	0.2900	−2.3969	−0.0532
HSi	271.4	102.0	2.8634	1.4566	0.1171	+0.6903	1.7362	0.3003	−1.9766	+0.0599
HP	259.6	77.7	2.6655	1.311	0.1670	+0.4858	1.3047	0.1887	−1.2949	+0.1985
HS	251.4	52.8	2.5134	0.9044	0.2175	+0.0342	0.5100	−0.0713	−0.085	+0.353
HCl	241.6	39.4	2.3928	0.7071	0.2490	−0.2420	0.0360	−0.1253	0.5791	+0.4898

[a]$v(A)$ and $v(H)$ are the atomic volumes of A and H, R_e is the equilibrium bond length (au), $r_b(H)$ the bonding radii of H and A (au), ρ_b is the bond critical point density (au), $q(A)$ is the charge on A, $M(A)$ and $M(H)$ are the intra-atomic dipole moments, M_{CT} is the charge transfer dipole moment, and M_{mol} is the total dipole molecule, which is the sum of $M(A)$, $M(H)$, and M_{CT}.

All calculations were performed at the B3LYP/6-311+G(2d,2p)//HF/6-31(d) level.

tronic charge along the bond path. Since the magnitude and the form of the electron density distribution vary along a bond, it is necessary to choose one particular point along the bond as characteristic of the bond. The most obvious point to choose is the bond critical point, a special and well-defined point that lies on the interatomic surface between the two atoms. There are three properties of the electron density at the bond critical point that provide us with useful information about the nature of the bond.

1. The value of the electron density at this point, the **bond critical point density** ρ_b.
2. The shape of the electron density distribution in a plane through the bond critical point and perpendicular to the bond as measured by its **ellipticity** ϵ.
3. The position of the bond critical point. The distance of this point from each of the nuclei is a measure of the size of each atom, that is, its **bonding radius** r_b.

Next, we illustrate these properties by discussing them together with some of the atomic properties for the diatomic hydrides of the elements of periods 2 and 3.

◆ 6.9 The Diatomic Hydrides of Periods 2 and 3

6.9.1 Bond Critical Point Density

The bond critical point density is a measure of the amount of electronic charge accumulated at this point. It seems reasonable to assume that this reflects the amount of density shared between the two atoms. In classical terms, a covalent bond is associated with a pair of shared electrons, so we could consider that the bond critical point density is a measure of the "covalent" character of the bond. However, we should be cautious about using this term because it cannot be clearly defined. Table 6.2 gives the values of ρ_b for the diatomic hydrides of periods 2 and 3. We see that the values of ρ_b for LiH and NaH are quite small and the values for the other hydrides increase steadily down each period, suggesting that the covalent character of the bond increases down each period. The atomic charges are also given in this table. We see that they decrease from LiH and from NaH to very small values for CH and SH, which are traditionally considered to have covalent bonds with very little polarity. These small charges are consistent with the electronegativies of these elements (C, 2.5; S, 2.4; H, 2.2). It is customary to associate the atomic charges with the "ionic " character or polarity of a bond, but again this quantity cannot be clearly defined and cannot therefore be measured. In the series LiH to HF, the charges reverse at CH and then increase up to HF. Similarly, the charges are reversed in HCl and are larger than in SH. But ρ_b continues to increase right through both series even though the bonds become more "ionic" or polar, as indicated by the increasing atomic charges. It is usually assumed that a bond becomes less covalent as it becomes more polar. However, this is clearly not the case if we take ρ_b to be a measure of the covalent character of the bond. We have to conclude that a bond can be both strongly covalent and very polar. But we must be aware that these are terms that cannot be clearly defined. In contrast, the analysis of the electron density according to AIM leads to two quantites that can be precisely defined and measured, namely, ρ_b, the bond critical point density, and the atomic charge. We further discuss these quantities and the ionic–covalent description of bonds in Chapters 8 and 9.

Values of ρ_b for ethane, ethene and ethyne, which are 0.249, 0.356, and 0.427 respectively, increase along this series consistent with the increasing bond order. However, ρ_b does not increase proportionally with the bond order, as we might at first expect. One obvious reason for this is that as the number of bonding electrons increases from a single to a double to a triple bond, the increased repulsion between the large number of electrons spreads the bonding density over a larger volume.

6.9.2 Bond Density Ellipticity

As we have seen, the electron density has a circular distribution around the axis of a diatomic molecule and generally around any single bond—for example, as in ethane. However, in ethene the electron density has an elliptical distribution around the CC bond (Figure 6.17), having a larger value in a direction perpendicular to the molecular plane than in the molecular plane. The magnitude of this ellipticity, which can be measured by the quantity ϵ, described in Box 6.2. It has a value of zero for ethane and any bond with a circular electron density distribution, but a value of 0.30 in ethene at the bond critical point. The ellipticity of the CC bonds in benzene is 0.23, which is consistent with the bonds having a considerable amount of double-bond character as suggested, for example, by the usual resonance structures. The elliptical nature of the electron density distribution of the CC bond in ethene is predicted by the classical bent-bond and VSEPR models, as well as by the σ–π model (Chapter 3).

For planar molecules such as ethene and benzene, the ellipticity ϵ can be considered to be a measure of the π character of a C—C bond, but this is not generally the case. The ellipticity of the triple bond in ethyne is zero, as expected for a linear molecule, even though in the σ—π model ethyne is considered to have two π bonds (Chapter 3). This is because the sum of the charge distributions of the electrons in the two π-bond orbitals is cylindrically symmetric, which together with the charge distribution of the electrons in the σ orbital gives an overall cylindrically symmetric charge distribution. We must remember that the

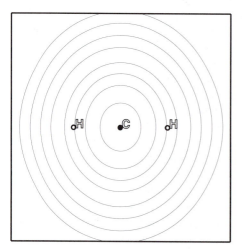

Figure 6.17 Contour map of ρ in the interatomic surface associated with the CC bond critical point in ethene. The plane of the plot is perpendicular to the molecular plane. The C and two H nuclei are projected onto the plane of the plot to indicate the orientation of the molecule. We see that electronic charge is preferentially accumulated in the direction perpendicular to the molecular plane, giving an elliptical shape to the electron density in this plane.

electron density distribution is a fundamental measurable property of a molecule, while the $\sigma-\pi$ model, the bent-bond model, and the VSEPR model are just models that may be used to interpret this distribution. Conversely, these models enable us to predict the electron density distribution in a qualitative way.

6.9.3 Bonding Radius, r_b

The distance from the bond critical point to each of the adjacent nuclei is a measure of the size of each atom in the bonding direction, and so we call it the **bonding radius, r_b.** This radius is the same as the conventional covalent radius in the case of a homonuclear diatomic molecule but not for any other molecule. The covalent radius of an atom is assumed to be constant from molecule to molecule, whereas the bonding radius changes, as we can see from the data for the period 2 and 3 hydrides in Table 6.2. The size of free neutral atoms decreases across any period with increasing effective nuclear charge. The size of an atom in a molecule also decreases with increasing positive atomic charge and conversely increases with increasing magnitude of the negative atomic charge. We see that the bonding radius of the hydrogen atom decreases across both periods as the magnitude of its negative charge decreases and then its positive charge increases.

The bonding radius of A in a diatomic hydride AH at first decreases across a period and then increases because there is a competition between two opposing effects:

1. The nuclear charge increases across the period, which decrease the size of the atom and therefore its bonding radius.
2. As its nuclear charge increases, A becomes more electronegative, thus attracting more electron density from the hydrogen, with the result that its overall positive charge decreases and eventually becomes negative, which increases its size and bonding radius.

On the left of the period, effect 1 predominates, while on the right effect 2 predominates, so that overall the bonding radius at first decreases and then increases again. However, the bond length R decreases across the period because of the large decrease in the bonding radius of hydrogen.

The changes in the size of an atom are shown both by the changes in the bonding radius and the nonbonding radius. These changes in atomic size can be clearly seen in the contour maps of the diatomic hydrides given in Figures 6.18 and 6.19. In LiH and NaH there is a large diffuse electron density on hydrogen and a smaller, much more tightly bound density on the Li or Na atom. The almost spherical nature of the contours and the large atomic charges in these molecules indicate their strongly ionic nature. In LiH, which can be represented approximately by $Li^+ H^-$, the contours on Li are much more closely spaced than those on H, even though the atoms have close to two electrons each, reflecting the contraction of the electron density as a consequence of the Li nuclear charge of $+3$, compared to the H nuclear charge of $+1$. Proceeding across the table, the electron density around the A atom becomes increasingly contracted and the atoms become smaller, as discussed earlier.

The large charge transfer from one atom to the other in the formation of a predominately ionic molecule is accompanied by a polarization of the electronic charge of the atoms in a

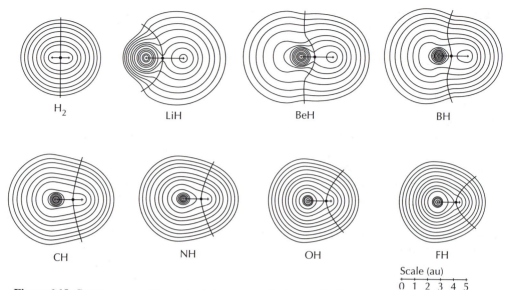

Figure 6.18 Contour maps of the ground state electronic charge distributions for the period 2 diatomic hydrides (including H_2) showing the positions of the interatomic surfaces. The outer density contour in these plots is 0.001 au. (Reproduced with permission from Bader [1990].)

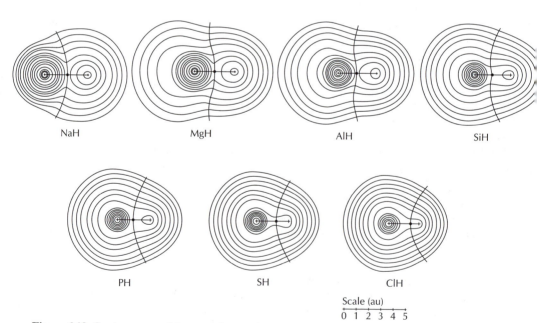

Figure 6.19 Contour maps of the ground state electronic charge distributions for the period 3 diatomic hydrides showing the interatomic surfaces. (Reproduced with permission from Bader [1990].)

direction counter to that of the charge transfer. The electronic charge distribution of the negative ion is polarized toward the cation, while the charge distribution of the cation is polarized away from the anion. These polarizations of the electron density are a necessary consequence of charge transfer. They provide the force that opposes the attractive force between the oppositely charged atoms, thus preserving equilibrium. The polarization of the cation is very small and not very evident in LiH and NaH because it occurs in the tightly held charge distribution of what is essentially the core. It is much more evident for Be and B in BeH and BH (Figure 6.18), because they have, respectively, one and two nonbonding electrons in the valence shell. We can think of this density situated on the side of the atom opposite from the bond as the nonbonding density. The atomic dipole moments $\mathbf{M}(A)$ and $\mathbf{M}(H)$, which are given in Table 6.2, are a measure of these polarizations of the electron density. The sum of the atomic moments and the charge transfer moment $\mathbf{M}_{CT}$ gives the total molecular dipole moment $\mathbf{M}_{mol}$.

◆ 6.10 Summary

We have seen in this chapter how the analysis of the electron density distribution of a molecule enables us to clarify and quantify some of the basic concepts relating to bonding. We can put the concept of an atom as it exists in a molecule on a quantitative basis The atom is defined by its interatomic surfaces and the $\rho = 0.001$ au contour on its nonbonded side, that is, as the union of its nucleus with its atomic basin. Then we showed how we can evaluate all the properties of these atoms by the integration of the appropriate property density over the atom. As examples, we discussed the atomic volume, the atomic charge, and the atomic dipole moment and showed that these properties are additive to give the value of the property for the molecule. Then we showed how we can identify the bonds in a molecule with the bond paths: lines of increased electron density between atoms, seen as ridges of electron density in a contour or relief map of a molecular plane containing the bond. The properties of a bond can be expressed in terms of the properties of the electron density at the bond critical point, in particular, the electron density at the bond critical point ρ_b, the position of the bond critical point that is a measure of the bonding radius r_b of an atom, and the ellipticity ϵ of the electron density, at this point. The bond critical point, electron density, and atomic charges enable us to put the somewhat vague concepts of ionic and covalent character on a more quantitative basis. We should note, however, that although we commonly speak of atoms and bonds as separate entities, the AIM analysis partitions a molecule into its constituent atoms, and there is no separate density than can be assigned to the bonds. The bonds appear only as a feature of the electron density of each atom, i.e., the concentration of density along the bond path connecting the two atoms.

In the following chapter we show how the topology of an important function of ρ, the Laplacian, enables us to obtain additional information from the analysis of the electron density distribution.

▶ Further Reading

R. F. W. Bader, *Atoms in Molecules: A Quantum Theory,* 1990, Oxford University Press, New York.
The definitive and authoritative book on AIM.
P. L. A. Popelier, *Atoms in Molecules: An Introduction*, 2000, Pearson Education, Harlow,
A simpler introduction to AIM than Bader's book.

▶ References

These papers discuss the AIM theory and its applications.

R. F. W. Bader, *Acc. Chem. Res. 18,* 9, 1985.
R. F. W. Bader, *Chem. Rev. 91,* 893, 1991.
R. F. W. Bader, R. J. Gillespie, and P. J. MacDougall, *J. Am. Chem. Soc. 110,* 7329, 1988a.
R. F. W. Bader, P. J. Macdougall, and C. D. H. Lau, *J. Am. Chem. Soc. 106,* 1594, 1988b.
R. F. W. Bader, R. J. Gillespie, and P. J. MacDougall, in *Molecular Structure and Energetics* Vol. 11
(J. F. Liebman and A. Greenberg, eds.), 1989, VCH, New York.
R. F. W. Bader, P. L. A. Popelier, and C. Chang, *J. Mol. Struct. (THEOCHEM) 255,* 145, 1992.
R. F. W. Bader, S. Johnson, T.-H. Tang, and P. L. A. Popelier, *J. Phys. Chem. 100,* 15398, 1996.

These papers discuss alternative methods of calculating atomic charges.

R. S. Mulliken, *J. Chem. Phys. 83,* 735, 1985.
P. Coppens, *X-ray Charge Densities and Chemical Bonding,* 1997, Oxford University Press, New York.
P. L. A. Popelier, F. M. Aiken, and S. E. O'Brien, *Royal Society of Chemistry Specialist Periodical
Report Vol 1*, Ed. A. Hinchliffe, 143 (2000).
A. E. Reed, R. B. Weinstock, and F. Weinhold, *J. Chem. Phys. 83,* 735, 1985.
A. E. Reed, F. Weinhold, and L. A. Curtiss, *Chem. Rev. 88,* 609, 1988.
K. B. Wiberg and P. R. Raben, *J. Comput. Chem. 14,* 1504, 1993.

▶ Appendix

The electron density distributions discussed in this book were obtained from wave functions generated with the program GAUSSIAN [Gaussian 94, revision B.1, M. J. Frisch, G. W. Trucks, H. B. Schlegel, P. M. W. Gill, B. G. Johnson, M. A. Robb, J. R. Cheeseman, T. Keith, G. A. Petersson, J. A. Montgomery, K. Raghavachari, M. A. Al-Laham, V. G. Zakrzewski, J. V. Ortiz, J. B. Foresman, J. Cioslowski, B. B. Stefanov, A. Nanayakkara, M. Challacombe, C. Y. Peng, P. Y. Ayala, W. Chen, M. W. Wong, J. L. Andres, E. S. Replogle, R. Gomperts, R. L. Martin, D. J. Fox, J. S. Binkley, D. J. Defrees, J. Baker , J. P. Stewart, M. Head-Gordon, C. Gonzalez, and J. A. Pople, Gaussian, Inc., Pittsburgh, 1995].

AIM data have been obtained with the commercial computer program MORPHY98, which is available from *http://www.ch.umist.ac.uk/morphy,* or with the AIMPAC suite of programs, developed by Bader's group at McMaster University (Canada), which can be obtained without charge from *http://www.chemistry.mcmaster.ca/aimpac.*

Since 1994 the commercial program GAUSSIAN (*http://www.gaussian.com*) has provided an option to perform a limited analysis of the electron density by the AIM method of AIM utilities. For this purpose, GAUSSIAN 98 is more reliable than GAUSSIAN 94.

7

THE LAPLACIAN OF
THE ELECTRON DENSITY

◼ ◼ ◼

◆ 7.1 Introduction

We have seen how we can extract useful information about the bonding in a molecule from the topological analysis of the electron density. However, we have not made much progress toward revealing the fundamental feature of chemical bonding postulated by Lewis, namely the electron pair. According to Lewis structures, there are bonding electron pairs in the valence shell of an atom in a molecule, and there are also nonbonding pairs or lone pairs in the valence shell of many of the atoms in a molecule. So far we have not seen any evidence for these electron pairs in our topological analysis of the electron density. We have seen that there is a concentration of electron density along a bond path, which arises from the increased probability of finding an electron in the region between the two bonded atoms. This increased probability is a consequence of the attraction exerted by a ligand on the electrons of the central atom. The Pauli principle allows only one electron of a given spin to be attracted close to a ligand and prevents all the other electrons of the same spin from crowding into this region; an electron of opposite spin is also allowed to be attracted into this region, however. So there is a higher probability of finding a pair of opposite-spin electrons in a bonding region than in other adjacent regions. Consequently, there is a concentration of electron density in this region. This increased probability of finding a pair of opposite-spin electrons in a bonding region and the consequent concentration of electron density in this region is the electron density equivalent of a Lewis bonding pair. But what evidence is there for lone pairs? If we look carefully at the electron density contour map for SCl_2 in Chapter 6 (Figure 6.4) we see slight bulges in the directions in which lone pairs on the sulfur atom are expected according to the VSEPR model. So it seems reasonable to associate these bulges with the expected lone pairs. In the next section we will see how more convincing evidence for bonding and nonbonding electron pairs can be obtained by studying an important function of the electron density, the Laplacian.

◆ 7.2   The Laplacian of the Electron Density

The **Laplacian** of the electron density, $\nabla^2 \rho$, is defined by

$$\nabla^2 \rho = \frac{\partial^2 \rho}{\partial x^2} + \frac{\partial^2 \rho}{\partial y^2} + \frac{\partial^2 \rho}{\partial z^2} \tag{7.1}$$

The Laplacian is constructed from second partial derivatives, so it is essentially a measure of the curvature of the function in three dimensions (Chapter 6). The Laplacian of any scalar field shows where the field is *locally concentrated* or *depleted*. The Laplacian has a negative value wherever the scalar field is locally concentrated and a positive value where it is locally depleted. The Laplacian of the electron density, ρ, shows where the electron density is locally concentrated or depleted. To understand this, we first look carefully at a one-dimensional function and its first and second derivatives.

Figure 7.1 shows a hypothetical monotonically decreasing function mimicking a one-dimensional electron density profile for a period 2 element. The value of the function $f(x)$

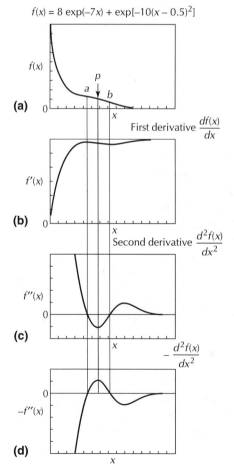

Figure 7.1 Plots of a monotonically decreasing function $f(x)$, its first and second derivative, and the negative of its second derivative. The slight "bulge" or shoulder in $f(x)$ is converted to a pronounced maximum in the negative of the second derivative $-f''(x)$.

decreases exponentially except for a very slight and hardly noticeable shoulder or bulge between points a and b on either side of point p. In the region between a and b, the value of $f(x)$ is greater than the *average* of its value at neighboring points. We say that the function is locally concentrated in this region. As expected for a monotonically decreasing function, its first derivative $f'(x)$ is everywhere negative and shows more features than $f(x)$ itself; in particular, it has a point of inflection at p. Its second derivative $f''(x)$ shows still more features and in particular has two zero values, which correspond to the two points of inflection a and b, at which the function changes from being concave up to concave down. Most importantly, the function $f''(x)$ is everywhere negative between a and b and has a minimum at the point p. We see that the nearly invisible slight bulge in $f(x)$ is much more evident as a negative region with a well-defined minimum in the second derivative $f''(x)$. Since we are interested particularly in regions of locally *increased* electron density, it is more convenient to consider the function $-f''(x) = -d^2f(x)/dx^2$. We see that the slight bulge in $f(x)$ then becomes a well-defined maximum in this function.

The function $-d^2f(x)/dx^2$ can be said to act as a kind of magnifying glass in that it converts the slight bulge in $f(x)$ to a large and well-defined maximum. The region between the two zero values of this function at a and b corresponds to the region in which the function $f(x)$ is locally concentrated. On either side of this region, $-d^2f(x)/dx^2$ is negative. These are the regions where $f(x)$ is locally depleted.

We can understand how $-d^2f(x)/dx^2$ shows us where $f(x)$ is locally concentrated or depleted in another way by approximating it by the finite difference formula:

$$-f''(x_0) = -\frac{d^2f(x)}{dx^2} \approx \frac{2}{(\Delta x)^2}\left[f(x_0) - \frac{f(x_0 + \Delta x) + f(x_0 - \Delta x)}{2}\right] \tag{7.2}$$

This approximation becomes more accurate the smaller the interval Δx. The term $\frac{1}{2}[f(x_0 + \Delta x) + f(x_0 - \Delta x)]$ is effectively the average of the function's values in the local neighborhood of x_0. As a result, $-f''(x)$ will be positive if $f(x)$ is greater than the values in its immediate neighborhood (since $(\Delta x)^2 > 0$). In other words, if the function is locally concentrated, then $-f''(x) > 0$. On the other hand, if $-f''(x) < 0$, the function is locally depleted.

The foregoing considerations carry over to three dimensions when a similar but more complicated finite difference formula is used. In particular, the value of $\rho(\mathbf{r})$ is greater than the average of its values over an infinitesimal sphere centered on $\mathbf{r}$ when $\nabla^2\rho$ is negative. Thus charge is locally concentrated in any region in which $\nabla^2\rho$ is negative and locally depleted in any region in which $\nabla^2\rho$ is positive. It is convenient, therefore, to define the function $L = -\nabla^2\rho$, because then L is positive in a region of local charge concentration and negative in a region of local charge depletion.

◆ 7.3 The Valence Shell Charge Concentration

As for the one-dimensional case, the function L makes features emerge from the electron density that ρ itself does not clearly show. What then does the function L reveal for the spherical electron density of a free atom? Because of the spherical symmetry, it suffices to focus on the radial dimension alone. Figure 7.2a shows the relief map of $\rho(r)$ in a plane through the nucleus of the argon atom. Figure 7.2b shows the relief map of $L(r)$ for the same plane, and Figure 7.2c the corresponding contour map. Since the electron density distribution is

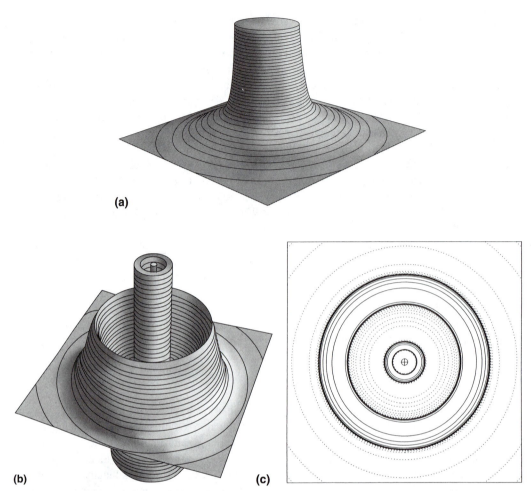

Figure 7.2 (a) Truncated version of the relief map of the electron density ρ for a plane containing the nucleus of the argon atom. (b) Relief map of $L = -\nabla^2 \rho$ for the same plane. (c) The corresponding contour map. In addition to the spikelike maximum in L at the nucleus, there are two shells of charge concentration. Outside each charge concentration there is a shell of charge depletion. Each successive pair of regions of charge concentration and charge depletion corresponds to one of the three quantum shells K, L, and M.

spherically symmetric, the same distribution is obtained for any plane through the nucleus. The electron density has a single maximum at the nucleus and decays steeply and monotonically along a radius from the nucleus, giving no indication of the existence of any electron shells. In contrast, $L(r)$ has a sharp inner spike centered on the nucleus surrounded by two more pairs of alternate maxima and minima along any radial line. These three pairs of maxima and minima correspond to the K, L, and M (or $n = 1, 2$, and, 3) shells. The outermost shell of charge concentration and the associated shell of charge depletion together constitute the valence shell.

Figure 7.3a is a radial plot of the electron density ρ, for an isolated sulfur atom. Although the ground state configuration $3s^2 3p_x^2 3p_y^1 3p_x^1$ is not spherical, for an isolated atom an average of the densities of all the possible configurations $3s^2 3p^4$ must be taken, giving an overall averaged density that is spherical. Indeed any isolated atom has a spherical electron density distribution. The radial density distribution has the same form as that for the argon atom, decreasing rapidly with increasing radial distance but showing no obvious feature corresponding to electron shells. Figure 7.3b shows the corresponding function L for the same interval. It shows the same three regions of charge concentration and charge depletion corresponding to the K, L, and M shells we saw for the argon atom. In discussing molecules, the topology of the outer (valence shell) charge concentration, the M shell in this case, is of particular interest to us. This charge concentration is called the **valence shell charge concentration** (VSCC). The radius of maximum charge concentration in the VSCC of sulfur is 1.34 au (70.9 pm). The radii of the spheres of maximum charge concentration for other elements are given in Table 7.1. It is clear from Table 7.1 that the radius of the maximum in the valence shell charge concentration decreases across the periodic table and increases down any group of the periodic table.

The inner shell charge concentrations of an atom in a molecule remain spherical, but the outer valence shell charge concentration is always distorted from a spherical shape, sometimes just a little, but sometimes so extensively that it breaks up into separate charge concentrations and sometimes, in very polar molecules, it disappears completely. In Figure 7.4a we show the contour map of L for SCl_2 in the molecular plane. In Figure 7.4b, an enlargement of the map of the sulfur atom, we see the circular cross sections of the unchanged spherical K and L inner shell charge concentrations, separated by regions of charge depletion. The valence shell charge concentration is, however, apparently broken up into three separate charge concentrations, each having a maximum. Two of these maxima are along the S—Cl bond paths so that the regions of charge concentration surrounding them are called **bonding charge concentrations** (bonding CCs). Figure 7.4c shows L in the symmetry plane perpendicular to the molecular plane. There is only a single region of charge concentration, with a gap in the vicinity of the point C and two maxima in this plane. The region of increased charge concentration surrounding each maximum is called a **nonbonding charge concentration** (nonbonding CC). The maxima of the two nonbonding CCs and the two bonding CCs have an approximately tetrahedral arrangement. The apparent third maximum, seen at D in Figure 7.4a, is not a real maximum but is the saddle point between the two maxima in the perpendicular plane shown in Figure 7.4c. For the sulfur atom, the valence shell charge concentration, which is spherical in the free atom, is perturbed in the molecule, hence it is no longer uniformly spherical but exhibits four maxima with an approximately tetrahedral arrangement. The valence shell charge concentration of each of the chlorine atoms in Figure 7.4c exhibits two maxima in this plane, although as we will see they are not two separate maxima in three dimensions but rather points on a ring of maximum charge of concentration surrounding each chorine atom.

The four maxima and the saddle point are *critical points* in the function $L(\mathbf{r})$ analogous to the maxima and saddle points in $\rho(\mathbf{r})$ discussed in Chapter 6. Every point on the sphere of maximum charge concentration of a spherical atom is a maximum in only one direction, namely, the radial direction. In any direction in a plane tangent to the sphere, the function L does not change; therefore the corresponding curvatures are zero. When an atom is part of

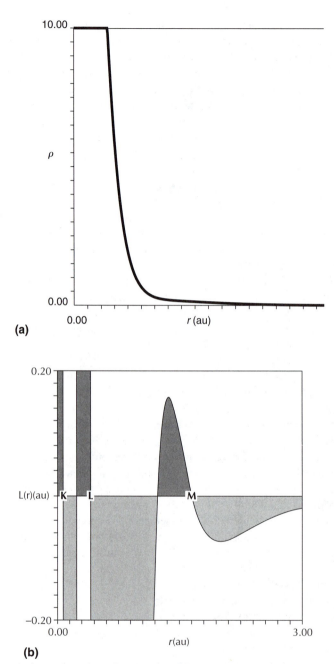

(a)

(b)

Figure 7.3 Truncated representation of ρ versus the distance from the nucleus for a spherically symmetric electron density of a free sulfur atom (^{3}P). (b) Truncated representation of L(r) at the same scale as (a). This function reveals the three shells K, L, and M constituting the sulfur atom. Each shell consists of a region of local charge concentration (dark areas) and a region of local charge depletion (light areas).

Table 7.1 Radii r_n (au) of Spheres of Maximum Charge Concentration for Quantum Shells $n = 2$ and 3

	Li	Be	B	C	N	O	F	Ne
r_2	2.49	1.59	1.19	0.94	0 .78	0.66	0.57	0.50

	Na	Mg	Al	Si	P	S	Cl	Ar
r_2	0.44	0.40	0.36	0.33	0 .30	0.28	0.26	0.24
r_3	3.44	2.55	2.08	1.76	1.52	1.34	1.20	1.08

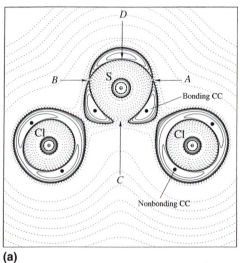

(a)

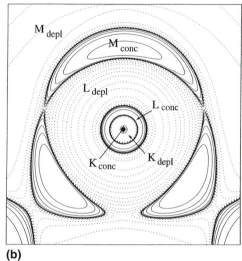

(b)

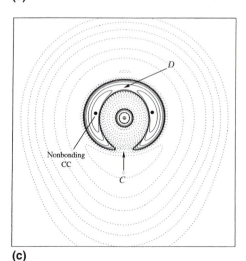

(c)

Figure 7.4 (a) Contour map of L in the molecular plane of SCl_2. The solid lines indicate positive L values and dashed lines negative values. The contours increase and decrease from a zero contour in steps $\pm 2 \times 10^n$, $\pm 4 \times 10^n$, $\pm 8 \times 10^n$, beginning with $n = -3$ and increasing in steps of unity. The maxima in each VSCC are indicated by dots. The labels A–D are referred to in the text. (b) Enlargement of L for the region of the sulfur atom showing the regions of charge concentration and depletion for each of the electron shells of the sulfur atom. (c) Contour map of L for the symmetry plane perpendicular to the molecular plane passing through the sulfur nucleus and bisecting the ClSCl angle; C and D refer to the same features shown in (a).

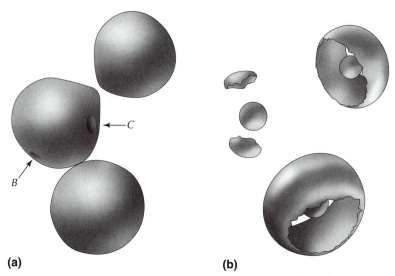

Figure 7.5 The three-dimensional isosurface (a) for $L(\mathbf{r}) = 0$ and (b) for $L(\mathbf{r}) = 0.60$ for SCl_2.

a molecule, its outer sphere of maximum charge concentration loses its spherical symmetry, so that in general it has a curvature in all three directions and now exhibits maxima and saddle points. The maxima are points at which all the curvatures in L are negative, and the saddle points are points at which two of the curvatures are negative and one is positive

We can gain more understanding of these plots of L by looking at some three-dimensional envelopes of constant L value. Figure 7.5a shows the constant density envelope for $L = 0$. This envelope shows that each valence shell charge concentration is distorted from a spherical shape but is not in fact broken up into separate regions of bonding and nonbonding charge concentrations. In this view of the $L = 0$ envelope, we see only the hole at C in the VSCC between the two bonding charge concentrations, of which we see a cross section in Figure 7.4a, and one of the two very small holes (A and B in Figure 7.4a), between the bonding CCs and the nonbonding CCs. Apart from these features, the $L = 0$ envelope of the sulfur VSCC is continuous although not truly spherical. We can see more detail in the $L = 0.60$ au envelope in Figure 7.5b, where the two nonbonding CCs in the VSCC of sulfur and the spherical inner core of each atom are separately visible. We also see clearly that the two apparent maxima in the VSCC of each chlorine atom in the two dimensional contour map of Figure 7.4b are points on a ring of maximum charge concentration in a torus of nonbonding charge concentration that surrounds each chlorine atom.

◆ 7.4 The Laplacian and the VSEPR Model

The two bonding maxima and the two nonbonding maxima in the valence shell charge concentration of the sulfur atom in SCl_2 have an approximately tetrahedral arrangement just like the two bonding domains and the two nonbonding domains of the VSEPR model. As we shall see, this correspondence is found for many other molecules, and so it seems reasonable to associate these regions of charge concentration with the bonding and nonbonding domains of the VSEPR model discussed in Chapter 4. We now look in more detail at the relationship between the valence shell charge concentrations and the VSEPR model.

We defined an electron pair domain in an approximate way as a region in which there is a high probability of finding an α-spin electron and a β-spin electron, that is, a pair of opposite-spin electrons. All the electrons in the valence shell of a free atom or monatomic ion are completely delocalized. Each electron may be thought of as having a domain that extends throughout the entire volume of the valence shell. Since these domains are completely overlapping and indistinguishable, all the valence shell electrons can be thought of as occupying the same spherical valence shell domain. An α-spin electron has the same probability of being found at any angle in the valence shell. Similarly, a β-spin electron has the same probability of being found at any angle in the valence shell. Same-spin electrons are kept apart by the operation of the Pauli principle and, to a lesser extent, by electrostatic repulsion. Thus the most probable relative angular arrangement of four same-spin electron is at the vertices of a tetrahedron. Since, however, this tetrahedron can have any orientation, the overall electron density distribution is spherical in any isolated atom.

In a molecule, the nucleus of a ligand X perturbs this distribution by attracting electrons in the valence shell of the central atom A. When an α-spin electron is attracted toward a ligand the Pauli principle ensures that four α-spin electrons will retain, at least approximately, their most probable tetrahedral arrangement. The β-spin electrons are similarly attracted by the ligands, and thus they adopt essentially the same most probable arrangement because the Pauli principle has no influence on the relative arrangement of opposite-spin electrons and does not prohibit opposite-spin electrons from being very close together or even in the same location. Consequently there is an enhanced probability of finding both an α-spin electron and a β-spin electron in each of the bonding regions and, because of the most probable tetrahedral arrangement of the two sets of opposite-spin electrons, also in certain nonbonding regions, namely, those that are in tetrahedral directions with respect to the bonding regions. The regions of an enhanced probability of finding an electron of α spin and electron of β spin are the regions defined in Chapter 4 as the electron pair domains of the VSEPR model. If electrons are found with a greater probability in certain regions of a molecule than in others, there must be a greater concentration of electron density in these regions. It is in these regions of greater concentration of electronic charge that we observe in the Laplacian of the electron density. Hence the Laplacian of ρ provides a physical justification for the domains of the VSEPR model—regions in which there is a high probability of finding a pair of electrons of opposite spin. In each of the regions where $L > 0$ there are one or more maxima, which are the points at which the concentration of charge is a maximum. These are accordingly the points at which there is a maximum probability of finding an α-spin electron and a β-spin electron—a pair of opposite-spin electrons.

We have seen that each chlorine atom in SCl_2 has a single toroidal nonbonding charge concentration rather than three separate nonbonding CCs corresponding to three nonbonding domains. This also corresponds to the VSEPR model because, as was pointed out by Linnett in his double-quartet model and as we discussed in Chapter 4, the Pauli principle does not lead to the formation of three nonbonding domains in this case. Rather the six nonbonding electrons have a maximum probability of being found anywhere in a ring perpendicular to the S—Cl bond axis and so have a toroidal domain corresponding to the toroidal charge concentration shown in Figure 7.5b. We discussed a very similar situation in Chapter 4 with respect to the HF molecule. A toroidal nonbonding domain and a corresponding toroidal charge concentration in L is a feature of all monatomic ligands.

The same correspondence between the domains of the VSEPR model and the charge concentrations in L is found in many other molecules, as we will now see.

7.4.1 The Water Molecule

Figure 7.6 shows contour plots of L for the water, ammonia, and methane molecules. In the plot of L for the molecular plane of the water molecule we see that the region of valence shell charge concentration extends over the whole molecule. The separate VSCCs are joined to each other in the bonding regions. In the VSCC of oxygen there are two bonding maxima and an apparent third maximum. This apparent maximum is, however, the saddle point between the two nonbonding maxima seen in the contour plot of L in the symmetry plane perpendicular to the molecular plane. These four maxima have an approximately tetrahedral arrangement. The two bonding maxima are the maxima of two bonding CCs that correspond to the two bonding domains of the VSEPR model. The two nonbonding maxima are the maxima of two nonbonding CCs corresponding to the VSEPR nonbonding domains. The angle of 138° between the maxima of the nonbonding CCs is larger than the angle of 103° be-

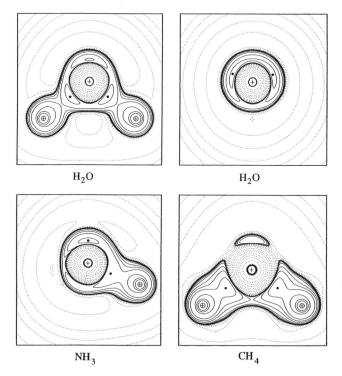

H_2O

H_2O

NH_3

CH_4

Figure 7.6 Contour maps of L for H_2O, (left; in the molecular plane; right, perpendicular to the molecular plane), NH_3 (in a symmetry plane through N and one H), and CH_4 (in a symmetry plane through C and two H's). The maxima in the valence shell charge concentration are indicated by the dots.

tween the bonding CCs. These angles are consistent with the domain model according to which the angular extension of a nonbonding domain is greater than that of a bonding domain, so that nonbonding domains subtend a greater angle at the nucleus than the bonding domains.

7.4.2 The Ammonia and Methane Molecules

In the contour plot of L for a plane through the nitrogen atom and one hydrogen atom of the ammonia molecule (Figure 7.6), we see one of the three bonding CCs in the valence shell CC of nitrogen and a nonbonding CC. The maxima of these four bonding CCs have an approximately tetrahedral arrangement. In the contour plot of L for a plane through the carbon and two hydrogen atoms of the methane molecule, also shown in Figure 7.6, we see two of the four bonding CCs in the VSCC of carbon. In this two-dimensional contour plot for CH_4, what appears to be a nonbonding maximum is not a maximum in three dimensions; rather, it is the saddle point between the two bonding charge concentrations that do not lie in this plane. The number and geometry of the CCs in the valence shell of the central atom in H_2O, NH_3, and CH_4 are the same as for the bonding and nonbonding domains of the VSEPR model.

7.4.3 The ClF₃ Molecule

According to the VSEPR model the T-shaped ClF_3 molecule has a trigonal bipyramidal arrangement of three bonding electron pairs and two nonbonding electron pairs in the valence shell of the central Cl atom.

$$F—\overset{..}{\underset{|}{Cl}}—F$$
$$F$$

Correspondingly, the ClF_3 molecule has three electron pair bonding domains and two nonbonding electron pair domains with an overall approximately trigonal bipyramidal geometry. This arrangement of electron pair domains is recovered by the arrangement of three bonding and two nonbonding charge concentrations in $L(\mathbf{r})$ of the valence shell of the Cl atom. Figure 7.7a plots L in the molecular plane. We see three bonding charge concentrations, each with a maximum in the corresponding bond path and an apparent fourth maximum, which is in fact the saddle point between the two nonbonding maxima on the chlorine atom (Figure 7.7b).

This exact correspondence between the number and geometrical arrangement of the electron pair domains of the VSEPR model and the number and geometrical arrangement of the maxima in L does not hold for all molecules. The valence shell charge concentrations always faithfully map the number and three-dimensional arrangement of the nonbonding or lone pair domains, as we have seen for SCl_2, H_2O, NH_3, and ClF_3, but this is not always true for bonding domains. In the Lewis and VSEPR models we consider that a single bonding pair of electrons is shared between two bonded atoms, which implies that the valence shells of the two bonded atom overlap or merge in the bonding region, leading us perhaps to expect to see only one bonding charge concentration. Although we do frequently observe just one bond-

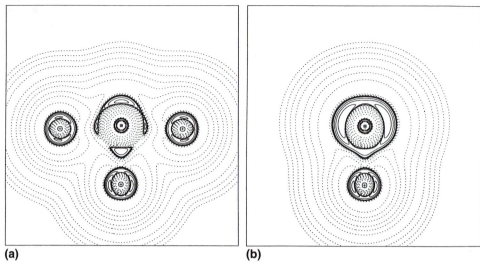

(a) **(b)**

Figure 7.7 Contour maps of L for ClF_3: (a) in the plane of the molecule and (b) in the symmetry plane through the equatorial fluorine atom.

ing charge concentration for each bond, sometimes we see two and sometimes no bonding charge concentration at all. Why is this?

We must remember that the valence shell charge concentration is not identical with the valence shell because there is also a region of valence shell charge depletion outside the valence shell charge concentration. In the formation of a bond, the two valence shells overlap either a little or extensively. The valence shell charge depletion regions therefore necessarily overlap, but the valence shell charge concentrations may not do so that there is then a separate VSCC for each atom. In such cases there may be a bonding maximum in each valence shell charge concentration, sometimes in only one of them, and occasionally in neither. In contrast, nonbonding electron pairs are always observed as nonbonding charge concentrations because nonbonding elecron pairs always remain in the valence shell of one atom and are not shared with any other atom. We next discuss when and under what conditions bonding charge concentrations are observed.

7.4.4 Predominantly Covalent Molecules

In a strong, short, predominately covalent bond between two atoms of equal or approximately equal electronegativity, as in ethane, C_2H_6 (Figure 7.8), the two valence shell charge concentrations merge with each other to give a single region of bonding charge concentration, usually with two maxima. These bonding maxima are situated equidistant from the bond critical point at the midpoint of the bond. In ethene, C_2H_4 (Figure 7.8), in which the CC bond is shorter than in ethane, the two bonding maxima move closer together. In ethyne, C_2H_2, the CC bond is still shorter, and the two maxima coalesce into a single bonding maximum. The situation is similar in the N_2 molecule (Figure 7.8), which also has a short bond and only one region of bonding charge concentration in which there are two maxima as in ethene.

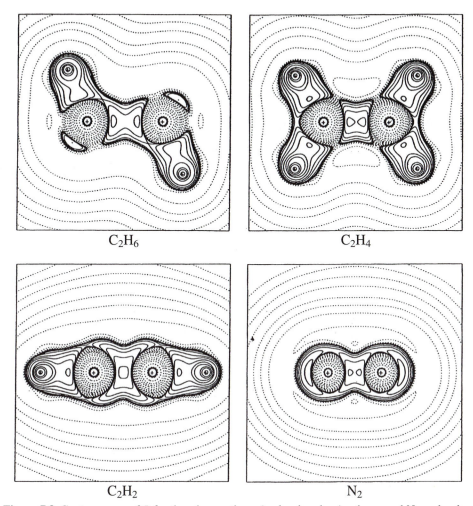

C_2H_6

C_2H_4

C_2H_2

N_2

Figure 7.8 Contour maps of L for the ethane, ethene (molecular plane), ethyne, and N_2 molecules.

Although according to the Lewis model, the CC double bond consists of two shared pairs and the CC triple bond in ethyne of three shared pairs, according to the VSEPR model there is only one bonding domain that has a greater electron density along the bond axis than along any other line between the two nuclei. This line of greater density is the bond path of the AIM theory. There is only one bond path between the carbon atoms in ethane, ethene, and ethyne, and it is along this bond path that the bonding charge concentrations are observed. Although there are two bonding maxima for the CC bond in ethene, these do not each correspond to one of the two electron pairs of the Lewis model. Rather, they are both associated with a single bonding domain.

In each of these hydrocarbons we also see a C—H bonding charge concentration with a maximum in the carbon valence shell corresponding to the C—H bonding domain, and in N_2 we see a charge concentration corresponding to the lone pair domain on each nitrogen atom. We note also that hydrogen is a unique ligand in that its valence shell and its core are identical, so that the C—H charge concentration also has a maximum at the hydrogen nucleus. This is the case for any A—H bond, as we can see in Figure 7.6 for H_2O, NH_3, and CH_4.

7.4.5 Predominately Ionic Molecules

Figure 7.9 gives contour plots of L for the molecules LiF, BeF_2, BF_3, and CF_4. When there is a large difference in the electronegativities of two bonded atoms, most of the electron den-

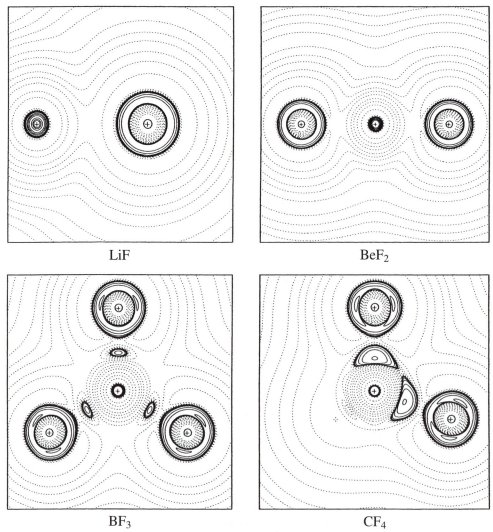

Figure 7.9 Contour maps of L for the LiF, BeF_2, BF_3, and CF_4 molecules.

sity in the valence shell of the less electronegative element is transferred to the valence shell of the more electronegative element. In the case of LiF this transfer is essentially complete, so the molecule consists of a cation-like lithium atom bound to an anion-like fluorine atom. No valence shell charge concentration is observed for the cation-like Li atom because it has lost most of its electron density, so there is no bonding maximum. We see only the VSCC for the essentially spherical core of this atom. The valence shell of the fluorine ligand is almost complete and therefore almost perfectly spherical, as we see in Figure 7.9, indicating that its valence shell electrons are not significantly localized into pairs. In this predominately ionic molecule in which, as we saw in Chapter 6, the bond critical point density is very small, no bonding CCs are observed. The situation is very similar in the molecule BeF_2. As the electronegativity of the central atom increases and the bond becomes shorter and more covalent, a small bonding charge concentration appears while a weak nonbonding toroidal CC also becomes evident on each fluorine atom. In CF_4, in which the bonds are more covalent and still shorter than in BF_3, four bonding charge concentrations with a tetrahedral arrangement are observed, two of which can be seen in the contour plot of the plane through two CH bonds (Figure 7.9). The nonbonding toroidal charge concentration of each fluorine atom becomes more evident from BF_3 to CF_4 as the increased electronegativity of the central atom increasingly localizes the nonbonding electrons of the fluorine ligands.

Figure 7.10 shows a contour plot of L for PF_3 in a plane through the phosphorus atom and one of the fluorine atoms. There are no bonding CCs in the valence shell of phosphorus, which is consistent with the large electronegativity difference between fluorine and phosphorus and the correspondingly large atomic charges (P, $+2.28$; F, -0.76), reflecting the large ionic character of the bonding. Although much of the valence shell density has been transferred to the ligands, the two nonbonding electrons remain in the phosphorus valence

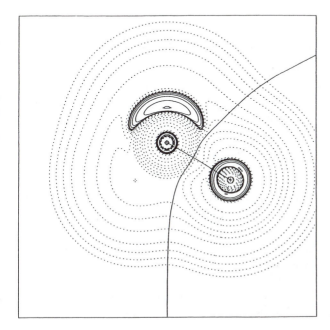

Figure 7.10 Contour map of L for the PF_3 molecule.

shell and are observed in the plot of L as a nonbonding CC. Because the nonbonding electrons are not removed from the valence shell by the ligands, charge concentrations corresponding to these lone pairs are always observed, as we saw for SCl_2, H_2O, NH_3, and ClF_3 and as we will see in many other examples in Chapters 8 and 9. These nonbonding or "lone pair" charge concentrations are generally more spread out around the core and are closer to the core and thus are consistent with the general shapes of the nonbonding and bonding domains of the VSEPR model. Summarizing, we can make the following statements.

- Maxima are always observed in the VSCC of an atom in a molecule corresponding in number and geometrical arrangement to the nonbonding electron pair domains of the VSEPR model.
- The nonbonding electrons of singly bonded monatomic ligands are not localized into pairs. Rather, they form a ring of six nonbonding electrons, which give rise to a toroidal charge concentration in the valence shell, consistent with the analogous toroidal domain of the VSEPR model.
- There is not always an exact correspondence between the maxima in the VSCC and the bonding domains of the VSEPR model. Whereas according to the VSEPR model a bond has a single bonding domain, there may be:
  1. A single bonding maximum in the VSCC of the more electronegative atom
  2. Two bonding maxima, one in each VSCC.
  3. One or two bonding maxima in a single shared VSCC in short covalent bonds
  4. No bonding maxima in the case of predominately ionic bonds

Further examples of contour maps of L are discussed in Chapters 8 and 9.

◆ 7.5 Electron Pair Localization and the Lewis and VSEPR Models

Electrons in the core of an atom are fully localized into spherical shells but not into opposite-spin pairs. In an isolated atom the valence shell electrons are similarly localized into a spherical shell. The Laplacian shows that in each of these spherical shells there is a spherical region of charge concentration and a spherical region of charge depletion. But in these regions there is no localization of electrons of opposite spin into pairs. There are no Lewis pairs or electron pair domains in an inner shell. The domain of each electron is spherical and fully delocalized through the shell.

In a molecule, the valence shells of the atoms are distorted from a spherical shape. In particular, the valence shell charge concentration that is made evident in the Laplacian of ρ develops maxima surrounded by regions of enhanced charge concentration in both the bonding and nonbonding regions. These regions of charge concentration show where there is an enhanced probability of finding both an α-spin electron and a β-spin electron, that is, a pair of opposite-spin electrons. There are no isolated electron pairs in an atom or a molecule, as one might imagine on the basis of a naive consideration of a Lewis model. Nor are there finite regions in which a pair of opposite-spin electrons will certainly be found. So the electron pair domains of the VSEPR model are not finite regions of space, as it is convenient and often useful to assume. There are only regions of charge concentration resulting from the enhanced probability of finding an electron of α spin and an electron of β spin in certain regions as a consequence of the perturbation produced by a bonded atom and the operation of the Pauli principle. A Lewis electron pair and a VSEPR electron pair domain are very approximate but useful models for these charge concentrations.

Or we may say that the charge concentrations of the Laplacian of the electron density provides the physical basis for these approximate models. Although electrons are not as localized into opposite-spin pairs in a molecule as the Lewis and VSEPR models assume, these models nevertheless have proved to be very useful over a long period of time. The importance of the Laplacian of ρ is that it reveals the limitations of these models and enables us to explain certain properties of molecules that are not consistent with the Lewis and VSEPR domain models. Lewis's genius was in recognizing the importance of electron pairs long before there was any physical evidence for them. But the Laplacian takes us further, showing us that electron pairs are not always present in molecules, and even when they are, they are not as localized as the approximate models may suggest. The usefulness of the Laplacian in providing a better understanding of bonding and molecular geometry will be evident again in Chapters 8 and 9.

The Laplacian of ρ has many applications, but in this chapter we have been able to give only a simple introduction to the Laplacian of ρ and to illustrate its usefulness in improving our understanding of molecular geometry.

◆ 7.6   Summary

The function L, which is the negative of the Laplacian (i.e., the second differential) of the electron density, shows where the electronic charge in an atom or molecule is either concentrated or depleted. Where $L > 0$, charge is concentrated; where $L < 0$, charge is depleted. Free spherical atoms have spherical regions of charge concentration and depletion corresponding to the electronic shells. The region of charge concentration corresponding to the outer or valence shell is called the valence shell charge concentration (VSCC). In a molecule, the valence shell charge concentration of an atom is perturbed, losing its spherical shape with the formation of maxima, between which there are saddle points. These maxima show where the electron density is most concentrated, while the saddle points between them indicate where it is less concentrated. These regions of greater concentrations of electronic charge are found in the bonding and nonbonding regions and are called bonding and nonbonding charge concentrations, respectively.

Nonbonding charge concentrations are observed in the valence shell charge concentration of all molecules that have lone pairs in the valence shell. Zero, one, or two bonding charge concentrations are observed for each bond path, depending on the relative electronegativities of the bonded atoms and on the bond length. When the difference in electronegativities is large and the bond is predominately ionic, most of the density of the central atom is transferred to the ligand. As a result, there is insufficient electron density to form a valence shell charge concentration in the less electronegative atom, while the ligand acquires an almost complete valence shell and therefore has an almost spherical VSCC in which no maxima are observed. With decreasing electronegativity difference, one and then two bonding CCs, one in each valence shell charge concentration, may be observed. In most predominately covalent molecules the two VSCCs merge to give a single shared VSCC in which there are two maxima. But in covalent molecules with very short bonds, such as the CC bond in ethyne, the two valence shell CCs merge to such an extent that only one maximum is observed at the midpoint of the bond. The Laplacian of ρ provides the physical basis for the Lewis and the VSEPR domain models and reveals the limitations of these models.

▶ Further Reading

These review papers discuss the AIM theory, including the Laplacian and its relationship to the VSEPR model.

R. F. W. Bader, *Atoms in Molecules: A Quantum Theory,* 1990, Oxford University Press, New York.
  The definitive and authoritative book on AIM.
R. F. W. Bader, P. J. Macdougall, and C. D. H. Lau, *J. Am. Chem. Soc. 106,* 1594, 1988b.
R. F. W. Bader, R. J. Gillespie, and P. J. MacDougall, in *Molecular Structure and Energetics* Vol. 11
  (J. F. Liebman and A. Greenberg, eds.), 1989, VCH, New York.
R. F. W. Bader, R. J. Gillespie, and P. J. MacDougall, *J. Am. Chem. Soc. 110,* 7329, 1988a.
R. F. W. Bader, S. Johnson, T.-H. Tang, and P. L. A. Popelier, *J. Phys. Chem. 100,* 15398, 1996.
P. Morse and H. Feshbach, *Methods of Theoretical Physics,* 1953, McGraw-Hill, New York.
  This book gives a fuller explanation of the Laplacian.
P. L. A. Popelier *An Introduction to the Theory of Atoms and Molecules,* 2000, Pearson Education,
  Harlow, U.K.
  A simpler introduction to AIM than Bader's book.

8

MOLECULES OF THE
ELEMENTS OF PERIOD 2

■ ■ ■

◆ 8.1   Introduction

In this chapter we discuss the bonding and geometry of the molecules of the elements of period 2. These molecules differ in several important ways from the those of the elements of period 3, which we discuss in the following chapter. One important difference is that the great majority of the molecules of the elements of period 2 have a coordination number of four or less. In contrast, the larger elements of period 3 and beyond can have coordination numbers of five, six, and even higher. It is convenient, therefore, to discuss the molecules of the period 2 elements first. Coordination numbers of greater than four for the period 2 elements are found only in infinite structures in the solid state and in cluster molecules, as explained in Box 8.1. In discrete molecules in which the ligands are not bonded together as in cluster molecules, the period 2 elements do not exceed a coordination number of four.

This chapter is based on the VSEPR and LCP models described in Chapters 4 and 5 and on the analysis of electron density distributions by the AIM theory discussed in Chapters 6 and 7. As we have seen, AIM gives us a method for obtaining the properties of atoms in molecules. Throughout the history of chemistry, as we have discussed in earlier chapters, most attention has been focused on the bonds rather than on the atoms in a molecule. In this chapter we will see how we can relate the properties of bonds, such as length and strength, to the quantities we can obtain from AIM.

◆ 8.2   The Relationship Between Bond Properties and the AIM Theory

As its name implies, AIM enables us to calculate such properties of atoms in a molecule as *atomic charge, atomic volume,* and *atomic dipole.* Indeed it shows us that the classical picture of a bond as an entity that is apparently independent of the atoms, like a Lewis bond line or a "stick" in a ball-and-stick model, is misleading. There are no bonds in molecules that are independent of the atoms. AIM identifies a bond as the line between two nuclei,

▲ BOX 8.1 ▼
Coordination Numbers Greater Than Four
for Period 2 Elements

Coordination numbers greater than four are found for some period 2 elements in non-molecular solids and in some cluster molecules. For example lithium in crystalline lithium chloride, which has the sodium chloride structure (Figure 1.7), has a coordination number of six. In this solid the bonding is predominately ionic and the coordination number of six is an exception to the radius ratio rules (Chapter 2), which are based on the concept of close packing of the anions around the lithium ion. In this structure it is the chloride ions that are close-packed and the lithium ion can be thought of as "rattling around" in a cage of chloride ions. There are also a number of predominately ionic carbides and nitrides such as Al_4C_3 and Ca_3N_2, which are crystalline solids in which the carbon or nitrogen atom is surrounded by an octahedron of metal atoms.

There are only a few discrete molecules in which a period 2 element has a coordination number greater than four. All are metal cage or cluster molecules such as $Co_6C(CO)_{13}^{2-}$, $Fe_6C(CO)_{16}^{2-}$, $Ru_6C(CO)_{17}$ (shown in the figure), and $Au_6C(P(C_6H_4Me)_3)_6^{2+}$, in which the carbon atom is encapsulated in an octahedral cage of metal atoms and so has a coordination number of six. There are also trigonal prism metal atom clusters that contain an encapsulated carbon atom such as $Co_6C(CO)_{15}^{2-}$. The coordination number of carbon in these molecules is dictated by the geometry of the atoms that surround it rather than by any property of the carbon atom itself. There are also examples of cluster molecules with other encapsulated main group atoms such as the trigonal prismatic $Co_6N(CO)_{15}^{-}$ and the octahedral $Co_6H(CO)_{15}^{-}$, as well as examples of metal clusters with five, seven, and more metal atoms containing encapsulated nonmetal atoms. The encapsulated atoms serve to provide the necessary num-

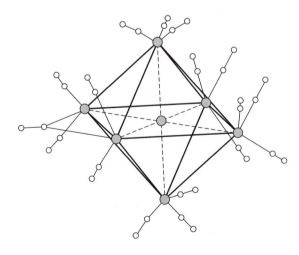

ber of electrons to give a stable cluster with a total number of electrons consistent with the rules developed by Wade and extended by Mingos. These rules, which also apply to some nonmetal clusters such as the boranes, have been partially justified by the molecular orbital approach. But this does not tell us anything about the nature of the bonding between the central atom and the cluster of metal atoms. The possibility of applying the VSEPR and LCP models and the AIM theory to cluster molecules is still largely a subject for future research.

called a *bond path,* that traverses the electron density distribution of each of the two bonded atoms and along which the electron density is concentrated. There is a unique point along the bond path at which it intersects the interatomic surface. This point, which we can consider as belonging to neither atom, or to both of them, is called the *bond critical point.* The electron density at this point, ρ_b, the *bond critical point density,* may be regarded as a characteristic property of the bond.

We assume that the bond critical point density is an approximate measure of the amount of density accumulated in the bonding region, that is, the amount of shared density. For bonds that are conventionally regarded as predominately covalent, ρ_b has a large value, and for bonds conventionally regarded as predominately ionic, ρ_b has a low value. In a hypothetical purely ionic bond, the value of ρ_b would be zero.

8.2.1 Bond Strengths and Lengths

The strength of a bond depends on the amount of electron density shared between the two bonded atoms, that is, on the value of ρ_b. However, this is clearly not the only factor determining bond strength, for otherwise a predominately ionic bond would have a very small or zero strength. In such a molecule, as in an ionic crystal, the attraction between the oppositely charged atoms is an important factor in determining the bond strength. And this must also be an important factor in any molecule in which the bonded atoms have opposite charges. Since the vast majority of bonds are polar, both factors must be taken into account in discussing the strengths of most bonds.

As we have seen in earlier chapters, an important and much discussed bond property is the *bond length.* The length of a bond depends on its strength, and it therefore also depends on the bond critical point density and on the atomic charges.

We saw in Chapter 5 that the length of a bond also depends on the coordination number of the atom to which it is bonded, increasing with increasing coordination number. So we can summarize the factors determining bond strengths and lengths as follows:

- The strength of a bond increases with increasing bond critical point density ρ_b and with increasing charges of the bonded atoms.

- The length of a bond decreases with increasing bond critical point density and increasing atomic charges, but increases with increasing coordination number of the atom to which it is bonded.

We can also discuss bond lengths in terms of the sizes of the bonded atoms. The distance along the bond path from the nucleus of an atom to the bond critical point is a measure of the size of the atom in the bonding direction, which we call the **bonding radius, r_b.** The bond length is the sum of the bonding radii of each of the bonded atoms. The bonding radius is a well-defined property of an atom in a molecule that can be obtained from the electron density distribution, in contrast to the arbitrarily defined covalent radius and ionic radius discussed in earlier chapters. It is applicable to any molecule, including molecules with very polar bonds, for which the concepts of ionic radii and covalent radii break down. Unlike the covalent and ionic radii, which are usually assumed to be constant from molecule to molecule, the bonding radius of an atom is not constant and is independent of the molecule in which it is situated.

- The length of a bond is equal to the sum of the bonding radii of the two bonded atoms.
- The bonding radius of an atom increases with increasing negative charge and decreases with increasing positive charge, just as we have seen for its ligand radius.

Clearly not all these atomic and bond properties are independent of each other and it can be difficult to disentangle one from another. Nevertheless we will find these properties useful for discussing the properties of molecules, as we do for some typical molecules of the period 2 elements in this chapter. In particular, the amount of accumulated or shared density, which we assume is approximately measured by the bond critical point density, represents what is commonly called the covalent contribution to the bonding. The atomic charges represent what is commonly called the ionic contribution.

◆ 8.3   The Nature of the Bonding in the Fluorides, Chlorides, and Hydrides of Li, Be, B, and C

8.3.1   Fluorides

Table 8.1 gives the experimental and calculated bond lengths and angles, the calculated atomic charges, the bonding radii of A and F, and the bond critical point density ρ_b for the fluorides of the period 2 elements, and some of their anions and cations. Contour maps of the electron density distributions for LiF, BeF_2, BF_3, and CF_4 are given in Figure 8.1.

Atomic Charges. The negative charge of the fluorine ligands decreases steadily from -0.92 in LiF to zero in F_2 as the electronegativity difference between the central atom and the ligand decreases (see Figure 8.2). The positive charge on the central atom increases with the increasing number of ligands from $+0.92$ in LiF to $+2.43$ in BF_3 and 2.45 in CF_4 and then decreases rapidly to zero in F_2 (Figure 8.2). We see from the electron density plots of LiF and BeF_2 that almost all the electron density is concentrated in an almost spherical region around each nucleus, so that each atom is very similar to an ion, consistent with the very ionic nature of these molecules indicated by the large atomic charges. From LiF to CF_4 an increasingly large fraction of the density is transferred toward the interatomic surface, thereby distorting the electron density distribution of each atom from its initially nearly spherical

Table 8.1 Molecular Parameters, Atomic Charges, and ρ_b Values for Period 2 Fluorides

Molecule	Bond Length, (pm)		Bond Angle, (°)			Atomic Change			Radii (pm)	
	Calc.	Exp.	Calc.	Exp.	ρ_b (au)	−q(F)	q(A)	q(A)q(F)	r_bF	r_bA
LiF	157.3	156.4			0.075	0.922	0.922	0.85	96.7	60.6
BeF_2	137.8	140	180	180	0.145	0.876	1.752	1.54	88.1	49.7
BF_3	131.4	130.7	120	120	0.217	0.808	2.433	1.96	86.9	44.5
CF_4	132.6	131.9	109.5	109.5	0.309	0.612	2.453	1.50	86.6	46.9
NF_3	138.2	138.5	101.9	102.3	0.314	0.277	0.834	0.68	74.9	63.3
OF_2	140.4	140.5	104	103.1	0.295	0.133	0.266	0.04	72.5	68.0
F_2	139.9	141.8	—	—	0.288	0	0	0	70.0	70.0
BeF_3^-	147.6	149	120	120	0.104	0.914	1.75			
BeF_4^{2-}	160.0	155.4	109.5	109.5	0.070	0.939	1.76			
BF_4^-	141.4	138.6	109.5	109.5	0.164	0.856	2.43			
CF_3^+	123.5	—	120	120	0.373	0.527	2.59			
NF_4^+	127.3	130	109.5	109.5	0.387	0.078	1.32			

shape and increasing ρ_b (Table 8.1). That the central atom loses a large part of its valence shell density in these molecules is clear from a comparison of the electron density distribution of BF_3, for example, with that of the free boron atom in Figure 8.1. The boron atom is much smaller in BF_3 than the free boron atom and is very similar in size to the B^{3+} ion.

Bond Lengths and Bonding Radii. As the negative charge on the fluorine atom decreases its size and in particular its bonding radius decreases (Figures 8.1 and 8.2). As the positive charge of the central atom A increases from LiF to CF_4, its bonding radius decreases correspondingly but then increases again to F_2 as the charge on A decreases rapidly to zero. The bond length, which is equal to the sum of the bonding radii, therefore decreases from 156.4 pm in LiF to 130.7 pm in BF_3 and 131.9 pm in CF_4 as both bonding radii decrease. Then the bond length increases to an almost constant value for NF_3, OF_2, and F_2 as the increase in the bonding radius of A roughly balances the decreasing bonding radius of fluorine (Figure 8.2).

Bond Critical Point Density. The value of ρ_b increases rapidly from LiF to CF_4 and then becomes essentially constant at a rather large value of approximately 0.3. We note that the change in ρ_b roughly parallells the change in the bond length, which decreases from LiF to CF_4, increases to NF_3, and then becomes essentially constant. The increasing bond critical point density from LiF to CF_4 indicates that the amount of shared density is increasing; that is, the bonds are becoming more covalent.

8.3.2 Polar Bonds and Ionic–Covalent Character

As discussed in Chapter 1, chemists have long recognized two types of bonds; ionic and covalent. However, a purely ionic bond is a hypothetical concept because in any bond there is

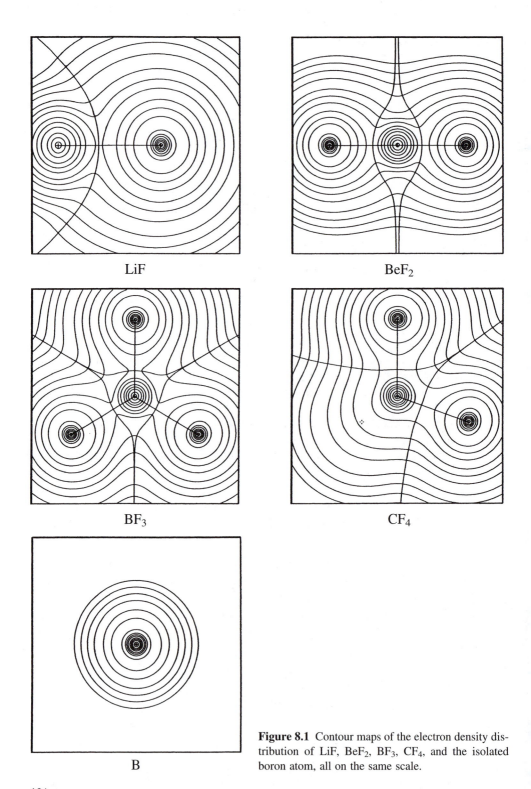

LiF

BeF$_2$

BF$_3$

CF$_4$

B

Figure 8.1 Contour maps of the electron density distribution of LiF, BeF$_2$, BF$_3$, CF$_4$, and the isolated boron atom, all on the same scale.

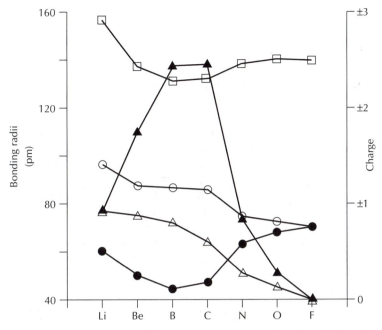

Figure 8.2 Atomic and bond properties of the period 2 fluorides: $\square$, bond length; $\bigcirc$, r_bF; $\bullet$, r_bA; $\blacktriangle$, qA; $\triangle$, qF.

at least some shared density; a purely covalent bond is rather rare because it is found only when the two bonded atoms are the same and have the same charge. In a purely ionic bond, the attraction between the two atoms is due only to their opposite charges, while in a pure covalent bond it is due only to the electron density accumulated between their nuclei. The vast majority of bonds have both characteristics. The bonded atoms have charges of opposite sign, and there is a certain amount of density accumulated between their nuclei. Thus the attraction between the two atoms is due both to the atomic charges and to the accumulated density. Such bonds are often said to have both ionic and covalent character, and it is usually assumed that if a bond is, say, 60% "ionic," it is 40% "covalent." This is not the case, however. Neither of these concepts can be clearly defined, and it cannot be assumed that there is a simple inverse relation between them. We have seen that from LiF to BF$_3$ the negative charge on fluorine decreases from -0.92 to -0.81, but the positive charge on A increases from $+0.92$ to $+2.43$. Is the A—F bond becoming less ionic, as would usually be assumed, or more ionic? At the same time, the amount of shared density, as indicated by the value of ρ_b, increases (Table 8.1), so it might be said that the bond is becoming more covalent. What is clear is that the attractive force between the two atoms increases because the charge on A increases much more than the charge on F decreases, and because the amount of accumulated density also increases. To say, as seems necessary, that the bond is becoming more ionic as well as more covalent is contrary to the usual usage of these terms, illustrating that the meaning of these poorly defined terms is not at all clear. Only the actual atomic charges and the bond critical point density can be given quantitative values and compared from one molecule to another. Nevertheless it is useful to be able to describe bonds in a qualitative manner, and so we will use the following descriptions:

- Bonds between atoms with large charges and a small value of the bond critical point density are described as predominately ionic.
- Bonds between atoms with very small charges and a large bond critical point density are described as predominately covalent.
- Many bonds have an intermediate character. They cannot be simply categorized as predominately ionic or predominately covalent and are described as polar.

The bonds in BF_3 and CF_4 are typical polar bonds. The large atomic charges and the large amount of electron density in the internuclear region make their bonds very strong. They have average bond enthalpies of 613 and 485 kJ mol^{-1}, respectively (Table 2.8). The CF bond is much stronger than a CC bond, which has an average bond enthalpy of 345 kJ mol^{-1} because the CC bond is nonpolar and does not have the additional strength resulting from the atomic charges. However, the CF bond is not as strong as the BF bond because the BF bond is shorter and the atomic charges are a little larger. The short length and great strength of the BF bond in BF_3 is adequately accounted for by the large atomic charges and high ρ_b values so there is no need to attribute it to back-bonding (Chapter 2).

Carbon tetrafluoride and all polyfluorinated hydrocarbons are very stable and inert molecules. This inertness can be attributed to the strength of the CF bonds and the close packing of the inert fluoride-like ligands around the carbon atom, which effectively prevent attack by a nucleophile on the carbon atom. In contrast, even though BF_3 has stronger bonds than CF_4, it is more reactive, forming adducts such as $BF_3 \cdot NH_3$ because the boron atom is only three-coordinated and there is space around it for an additional ligand.

8.3.3 Chlorides

Table 8.2 gives the experimental and calculated bond length and bond angles, and the calculated atomic charges, bonding radii, and bond critical point densities for the chlorides of the period 2 elements and some of their anions and cations. Plots of the electron density distributions for LiCl, $BeCl_2$, BCl_3, and CCl_4 are given in Figure 8.3.

Table 8.2 Molecular Parameters, Atomic Charges, and ρ_b Values for Period 2 Chlorides

Molecule	Bond Length (pm)		Bond Angle (°)		$\rho_b(au)$	Atomic Charge			Radii (pm)	
	Calc.	Exp.	Calc.	Exp.		q(Cl)	q(A)	q(A)q(Cl)	r_bA	r_bCl
LiCl	202.2	220.1	—	—	0.047	−0.91	+0.91	0.83	68.4	133.8
$BeCl_2$	179.8	—	180	180	0.097	−0.84	+1.68	1.41	56.8	123.0
BCl_3	175.0	174.2	120	120	0.157	−0.64	+1.93	1.24	53.7	121.1
CCl_4	179.7	177.1	109.5	109.5	0.182	−0.09	+0.35	0.03	81.8	97.1
NCl_3	179.1	175.9		107.1	0.176	+0.08	−0.24	0.09	87.3	91.9
OCl_2	172.8	170	112.8	111.2	0.184	+0.23	−0.46	0.11	87.7	85.1
FCl	166.5	162.8	—	—	0.187	+0.38	−0.38	0.14	87.8	78.0
BCl_4^-	188.1	183.3	109.5	109.5	0.122	−0.70	+1.81			
CCl_3^+	165.8	—	120	120	0.235	+0.22	+0.33			

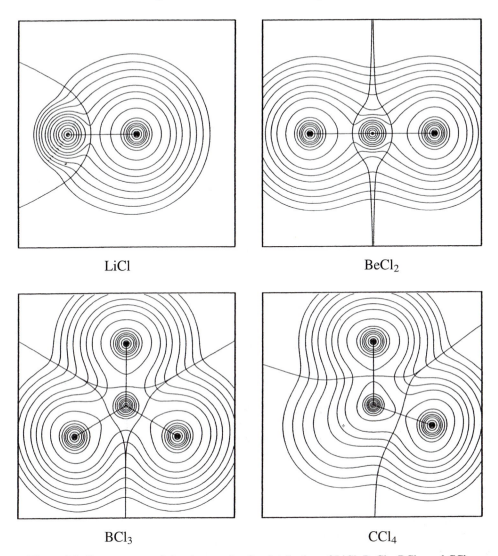

LiCl

BeCl$_2$

BCl$_3$

CCl$_4$

Figure 8.3 Contour maps of the electron density distribution of LiCl, BeCl$_2$, BCl$_3$, and CCl$_4$.

Atomic Charges. The negative charge of the Cl ligand decreases from LiCl to BCl$_3$ and to almost zero in CCl$_4$, consistent with the small electronegativity difference between Cl (2.8) and C (2.5) and then becomes positive in NCl$_3$, OCl$_2$, and FCl, since in these molecules the electronegativity of the central atom is greater than that of chlorine (Figure 8.4). The charge on the central atom A increases from Li to B, drops sharply to a small value for carbon, and then becomes increasingly positive. The bonds in CCl$_4$ have a very low polarity and are predominately covalent. The change from a predominately ionic bond in LiCl to a predominately covalent bond in CCl$_4$ is also indicated by the increasing deviation of the electron density distribution of the atoms from a spherical shape as can be seen in Figure 8.3.

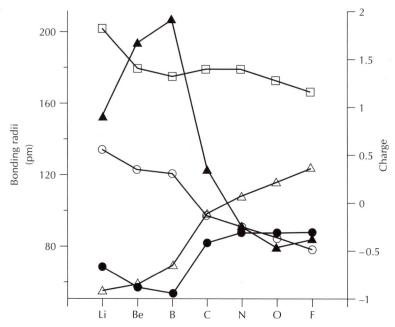

Figure 8.4 Atomic and bond properties of the period 2 chlorides: □, bondlength; ○, r_bCl; ●, r_bA; ▲, qA; △, qCl.

Bonding Radii and Bond Lengths. The bonding radius of chlorine decreases with decreasing negative charge and increasing positive charge from LiCl to FCl (Figure 8.4). The bonding radius of A decreases with increasing positive charge from LiF to BF_3 and then increases slowly as the charge on A changes from a small positive value to a slowly decreasing negative value. As a consequence, the bond length decreases to BF_3, increases to CF_4, then slowly decreases again. The A—Cl bond length therefore continues to decrease after CCl_4.

Bond Critical Point Density. The value of ρ_b increases from LiCl to CCl_4 and then becomes essentially constant in the following predominately covalent molecules. The bond critical point density is lower than in the corresponding fluorides, and the A—Cl bonds are longer than the A—F bonds because of the larger size of the Cl atom. It is a general observation that longer bonds have lower bond critical point densities. Longer bonds are therefore also generally weaker than shorter bonds. For example the B—Cl bond enthalpy is 456 kJ mol^{-1} compared to 613 kJ mol^{-1} for the shorter BF bond (Table 2.8). This difference in the strengths of the BF and BCl bonds has an important effect on the Lewis acid strengths of BF_3 and BCl_3 as we discuss in Box 8.2.

8.3.4 Hydrides

Table 8.3 gives the experimental and calculated bond length and angles, and the calculated atomic charges, bonding radii, and the bond critical point densities, for the hydrides of the

▲ BOX 8.2 ▼
The Lewis Acid Strengths of BF_3 and BCl_3

It has been known for long time that the Lewis acid strength of BCl_3 is greater than that of BF_3, although the calculated charge on boron is $+2.43$ in BF_3 but only $+1.93$ in BCl_3 (Tables 8.1 and 8.2). The commonly accepted explanation is that there is more back-bonding (Chapter 2) in BF_3 than in BCl_3 because the 2p orbital on fluorine is more comparable in size to the 2p orbital on boron than is the 3p orbital on chlorine. However, the LCP model provides an alternative and simpler explanation. On coordination to a base such as NH_3, the geometry of the BX_3 molecule changes from three-coordinated planar triangular to approximately tetrahedral and the BF and BCl bonds increase in length accordingly. The BF bond increases in length from 130.7 pm to 136.7 pm in $BF_3 \cdot NH_3$ and the BCl bond increases in length from 174.2 pm to 183.7 pm in $BCl_3.NH_3$ (Figure 1). The stretching of the B—X bonds in the formation of a complex is an endothermic process. So, because B—Cl bonds are weaker than B—F bonds, more energy is needed to stretch a BF bond than a BCl bond. The greater strength of a BF bond is shown by:

1. The average bond enthalpies, which are 613 kJ mol^{-1} for a BF bond and 456 kJ mol^{-1} for a BCl bond.
2. The larger atomic charges in BF_3 [$q(B) = +2.43$, $q(F) = -0.81$] than in BCl_3 [$q(B) = +1.93$, $q(Cl) = -0.64$].
3. The larger ρ_b value of 0.217 in BF_3, compared to the value of 0.157 in BCl_3

Since less energy is needed to stretch a BCl bond than a BF bond in forming a complex with NH_3, the energy gained in the formation of the BN bond is offset less by the energy needed to stretch a BCl bond than by the energy needed to stretch a BF bond. So the energy of formation of a BF_3 adduct is less than that of the corresponding BCl_3 adduct, and consequently BCl_3 is a stronger Lewis acid than BF_3 (Rowsell et al.)

In contrast, with very weak bases such as CO and MeCN, BF_3 forms stronger complexes than BCl_3. In these weak complexes $BX_3 \cdot CO$ and $BX_3 \cdot NCMe$ the BC and BN bonds are much longer, the BX_3 molecule is not significantly distorted from planarity, and the BX bond lengths remain almost unchanged as we can see in the table where $BX_3 \cdot CO$ and $BX_3 \cdot NH_3$ are compared. The relative stabilities of these complexes depends only on the charge on boron because no significant stretching of the BX bond is involved. Hence BF_3 forms stronger complexes with CO and other weak bases than BCl_3 and so is the stronger Lewis acid. We note also that the BC bond in $BF_3 \cdot CO$ and $BCl_3 \cdot CO$ is much longer than the BN bond in the corresponding NH_3 complexes confirming that the BF_3 and BCl_3 complexes with CO are much weaker than their complexes with NH_3.

Table Box 8.2 Bond Lengths and Bond Angles in BF_3 and BCl_3 Complexes

	B—X (pm)	XBX (°)	B—C(N) (pm)
BF_3	131.0	120	—
$BF_3 \cdot CO$	131.1[a]	120	288.6
$BF_3 \cdot NH_3$	138.0	111.0	171.2
BCl_3	174.2	120	—
$BCl_3 \cdot CO$	174.4	120	322.1
$BCl_3 \cdot NH_3$[a]	183.7	113.5	161.0

[a]Ab initio calculated values.

Table 8.3 Geometrical Parameters, Atomic Charges, and ρ_b Values for the Hydrides of the Period 2 Elements

Molecule	Bond Lengths (pm)		Bond angles (°)		Atomic Charges			Radii (pm)	
	Exp.	Calc.	Exp.	Calc.	q(H)	q(A)	$\rho_b(au)$	r_bH	r_bA
LiH	—	159.2	—	—	−0.91	+0.91	0.041	88.0	71.2
BeH_2	—	132.6	180	180	−0.87	+1.73	0.101	74.8	57.8
BH_3	—	118.9	120	120	−0.70	+2.11	0.189	65.9	53.0
CH_4	109.5	109.0	109.5	109.5	−0.04	+0.18	0.274	39.0	70.0
NH_3	101.5	101.6	107.2	107.9	+0.35	−1.05	0.332	28.1	73.5
OH_2	95.8	96.3	104.5	106.3	+0.63	−1.25	0.362	20.5	75.8
FH	91.8	92.4	—	—	+0.78	−0.78	0.365	15.5	76.9
BH_4^-	123.7	123.7	109.5	109.5	−0.67	+1.69			
NH_4^+	103.2	102.6	109.5	109.5	+0.48	−0.93			
CH_3^+	—	109.3	120	120	+0.28	+0.16			

period 2 elements. Contour plots of their electron density distributions are shown in Figure 8.5. Similar calculated data for the period 2 diatomic hydrides, most of which have not been experimentally observed, were given in Table 6.2 and Figure 6.18.

Atomic Charges. The atomic charges in LiH, BeH$_2$, and BH$_3$ (Figure 8.6) are large, indicating that these are very polar molecules, but both the charges decrease to very small values in CH$_4$. Thus the bonds in methane have a very low polarity and are predominately covalent, consistent with the very similar electronegativities of carbon and hydrogen. Although the electronegativity of hydrogen is a little smaller than that of carbon, the hydrogen ligands

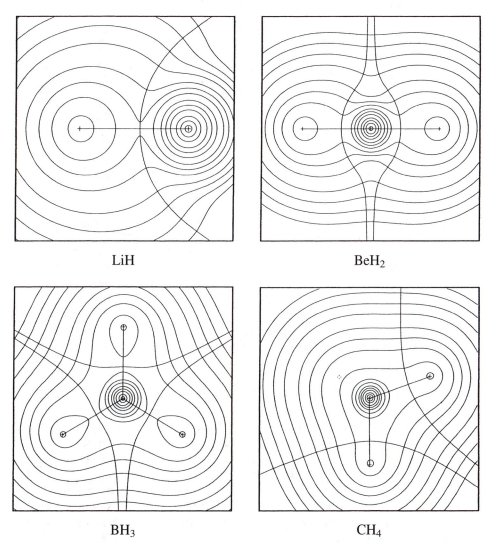

LiH

BeH$_2$

BH$_3$

CH$_4$

Figure 8.5 Contour maps of the electron density distribution of LiH, BeH$_2$, BH$_3$, and CH$_4$

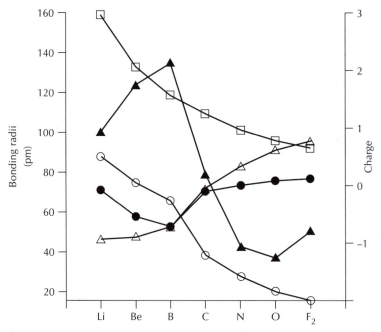

Figure 8.6 Atomic and bond properties of the period 3 hydrides: □, bond length; ○, r_bH; ●, r_bA; ▲, qH; △, qA.

nevertheless have a small negative charge, suggesting that the electronegativity of hydrogen should be assigned a slightly higher value than that of carbon. However, as we pointed out earlier, electronegativity is only an approximate quantity that cannot be considered to have a truly constant value from molecule to molecule. Moreover, this may be particularly true for hydrogen, which is a unique atom in that hydrogen has no inner core electrons and no nonbonding electrons, with the result that the electrons in its valence shell in a molecule act as both bonding and nonbonding electrons. The charge on hydrogen is positive from NH_3 to HF, as expected from the greater electronegativity of N, O, and F, and the bond polarity increases again, but in the reverse sense.

Bonding Radii and Bond Lengths. As the charge on hydrogen decreases from a large negative value in LiH to a very small value in CH_4 and then has an increasing positive value, there is a considerable decrease in the size of the hydrogen ligand, as can be clearly seen in Figure 8.6. The decrease in size of the hydrogen atom appears to be the most important factor determining the decrease in bond length from LiH to FH, more than compensating for the relatively small increase in the bonding radius of the central atom after BH_3. This large variation in the size of the hydrogen atom is a consequence of its lack of any core electrons. We have seen a relatively large decrease in the *ligand* radius of hydrogen from 110 pm when bonded to boron to 82 pm when bonded to nitrogen (Table 5.6). The *bonding* radius of hydrogen decreases even more sharply, from 65.9 pm in BH_3 to 28.1 pm in NH_3. These results also show that little confidence can be placed on the value of 120 pm for the van der Waals radius of hydrogen (Table 5.1), which must similarly vary greatly with its charge.

8.3.5 Beryllium and Boron Hydrides

Neither of the simple molecules BeH_2 or BH_3 is known. BeH_2 is a polymeric solid of unknown structure. There are many **boranes** (boron hydrides). The simplest is B_2H_6, which has a bridged structure with an approximately tetrahedral arrangement of the hydrogens around each boron. Contour plots of the electron density distribution are given in Figure 8.7a,b, and the calculated and experimental geometrical parameters are given in Figure 8.7c,d. The terminal ligand–ligand distances of 203 and 207 pm (Figure 8.6e) are close to the values for the calculated structures of BH_4^- (202 pm) and BH_3 (203 pm). The H···H distance between the two bridging hydrogens is, however, only 194 pm. It appears that because the negatively charged bridging hydrogen atoms are attracted directly toward each other by two positive boron atoms, their electron distributions are slightly compressed. This allows the opposite HBH angle to open up to 121.8°, increasing the H···H distance to 207 pm. This opening up of the angle between the terminal hydrogens in B_2H_6 is analogous to the opening up of the HCH angle in ethene as the two bonding pairs of electrons of the double bond are pulled toward each other by the attraction of the two positive carbon cores, merging to form a single four-electron domain as discussed in Chapter 4.

The description of the bonding in B_2H_6 has been the subject of discussion over many years because there are only 12 valence electrons, which is fewer than the 16 electrons needed for there to be two-electron bonds between each pair of adjacent atoms. B_2H_6 is a simple example of a general class of molecules that are called *electron deficient* because there are insufficient valence electrons to form a two-electron bond between each pair of bonded atoms. In view of the large atomic charges, this molecule can to a rough first approximation be considered to be composed of an approximately tetrahedral arrangement of anion-like hydrogen ligands around each of the two cation-like boron atoms (Figure 8.7f).

A Lewis-type covalent description can be given in terms of the two resonance structures shown in Figure 8.7g. Another useful model is to consider B_2H_6 as protonated $B_2H_4^{2-}$, which is isoelectronic with C_2H_4, using the bent-bond model of these molecules (Figure 8.7h). As is often the case, the formal charges in this model of B_2H_6 bear no relation to the real charges. From the point of view of the VSEPR model it can be considered that there is an approximately tetrahedral arrangement of four electron pairs around each boron atom. Two of these pairs are shared with the two terminal hydrogen ligands and the other two are shared with each of the bridging hydrogens (Figure 8.7i), forming what are called two-electron, three-center (2e,3c) bonds, as opposed to normal Lewis bonds, which may be described as 2e,2c bonds. A hybrid orbital model description uses four sp^3 hybrid orbitals on each boron atom (Figure 8.7j). Two of these sp^3 orbitals are used to form the two terminal BH bonds and the other two to form a bond with the bridging hydrogen, giving a three-center orbital extending over the bridging hydrogen and both boron atoms. This description does not account for the deviations of the bond angles from the tetrahedral value, although it could be improved by adjusting the ratio of the s and p contributions to the hybrid orbitals. But this is again a description of the bonding, not an explanation of the geometry. Despite their apparent dif-

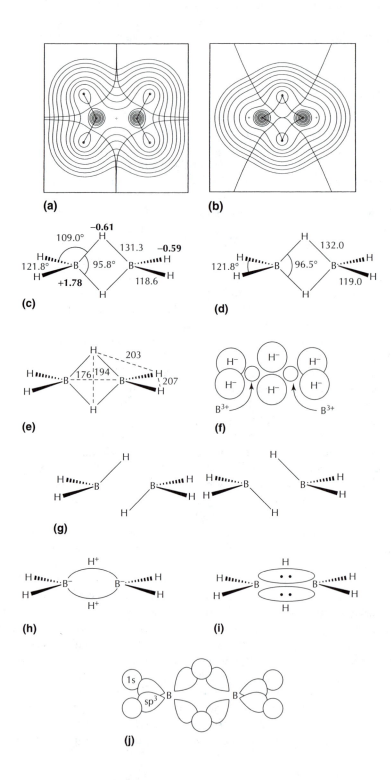

ferences these models give a very approximate description of the same electron density distribution in Figure 8.7a,b.

The contour maps of the electron density in Figure 8.7a,b show a bond path between each boron and the bridging hydrogens and between the boron atoms and the terminal hydrogens. However, the bond paths to the bridging hydrogens cannot be identified with single electron pair bonds; rather, the density of one electron pair is spread out over two boron atoms and one bridging hydrogen as the approximate models attempt to depict, and is concentrated along two bond paths, forming two rather weak long bonds. There is no bond path between the two boron atoms. There is, therefore, no bond between these two atoms, even though the boron–boron distance (176 pm) is equal to twice the covalent radius of boron (Table 2.1). But we must remember that because each boron is positively charged, its actual radius is considerably smaller than that of a neutral boron atom. We are reminded once again of the very approximate nature of covalent radii, which are often assumed to be constant from one molecule to another despite changes in the atomic charge.

The many higher boranes such as B_5H_9 and $B_6H_6^{2-}$ are similarly electron deficient and cannot be described by a single Lewis structure. They can often be described in terms of a combination of two- and three-center bonds. Alternatively, their structures can be rationalized by electron-counting schemes such as those proposed by Wade. Analysis of the electron density of these molecules by the AIM method shows that there are bond paths between all adjacent pairs of atoms. So from the point of view of the AIM theory there are bonds between each adjacent pair of atoms, but these cannot all be regarded as Lewis two-center, two-electron bonds as is the case in B_2H_6.

◆ 8.4 The Geometry of the Molecules of Be, B, and C

The hydrides, fluorides, and chlorides of Be, B, and C all have the expected AX_2, AX_3, and AX_4 geometries consistent with both the VSEPR and LCP models. The data in Table 8.4 remind us again of the importance of ligand close packing. Although the bond lengths increase from three- to four-coordination, the interligand distances remain nearly constant. Moreover, we can see from Tables 8.1–8.3 that the atomic charges in related molecules such BF_3 and BF_4^- are very similar, which means that the large increase in bond length with increasing coordination number in these molecules cannot be related to change in the atomic charges, that is, to the ionic contribution to the bonding.

The localization of the valence shell electrons of the central atom into pairs that is the basis of the VSEPR model can be clearly seen in the contour map of L for a plane through the carbon atom and two Cl ligands for the CCl_4 molecule given in Figure 8.8. There is a maximum in the valence shell charge concentration of both carbon and chlorine along each

---

Figure 8.7 Diborane, B_2H_6. (a) Contour map of ρ_b in the plane of the terminal hydrogens. (b) Contour map of ρ_b in the plane of the bridging hydrogens. (c) Calculated geometry. (d) Experimental geometry. (e) Interatomic H···H distances. (f) Ionic model. (g) Resonance structures. (h) Protonated double-bond model. (i) VSEPR domain model showing the two three-center, two-electron bridging domains. (j) Hybrid orbital model.

Table 8.4 Bond Length and Coordination Number in the Chlorides, Fluorides, and Hydrides of Boron and Carbon

	Fluorides			Chlorides			Hydrides	
Molecule	Bond Length (pm)	F···F (pm)	Molecule	Bond Length (pm)	Cl···Cl (pm)	Molecule	Bond Length (pm)	H···H (pm)
BF_3	130.7	226	BCl_3	174.2	302	BH_3^a	118.7	206
BF_4^-	138.6	226	BCl_4^-	183.3	299	BH_4^-	123.7	202
CF_3^{+a}	123.5	214	CCl_3^{+a}	165.8	287	CH_3^{+a}	109.3	189
CF_4	131.9	215	CCl_4	176.7	289	CH_4	109.5	178

[a]Calculated data.

of the two CCl bonds in this plane. The apparent third maximum in the carbon VSCC on the opposite side from the other two maxima is in fact a saddle point between the two maxima along each of the two bonds that do not lie in this plane. The four maxima in the carbon VSCC have a tetrahedral geometry, as expected. In the valence shell charge concentration of each chlorine ligand in CCl_4 in the plane plotted in Figure 8.8 there appear to be two maxima in addition to the bonding maximum. These are actually just cross-sections of a toroidal CC that surrounds the chlorine atom that is very similar to the toroidal CC surrounding each Cl atom in SCl_2, illustrated in Figure 6.3a. This CC results from the six nonbonding electrons (see Chapters 4 and 7). The plots of L for BCl_3, $BeCl_2$, and LiCl in Figure 8 illustrate the gradual transition from predominately covalent bonding in CCl_4 to predominately ionic bonding in LiCl.

The tetrahedral AX_4 geometry is common for both Be and B, and there are many examples, in addition to anions such as BF_4^-. For example, Lewis acid–base complexes of BF_3 and BCl_3 with ammonia and many other bases and molecules such as the molecule $Cl_2Be(Et_2O)_2$, and $BeCl_2$ in the solid state, which has a polymeric chain structure (Figure 8.9), all have a tetrahedral geometry around the central atom. And, of course, much of organic chemistry is based on the tetrahedral carbon atom.

◆ 8.5 Hydroxo and Related Molecules of Be, B, and C

Table 8.5 gives the bond lengths and the calculated atomic charges and bond critical point densities for some hydroxo molecules of Be, B, and C. Figure 8.10 gives the geometrical parameters and electron density contour map for $B(OH)_3$ which, as in the case of BF_3, clearly shows the transfer of a large amount of the valence shell density from the boron atom to the hydroxide ligands. The values for the atomic charges of the central atom and the OH group are very similar to, but slightly smaller than, the atomic charges for the corresponding fluorides, consistent with the smaller electronegativity of O versus that of F. For example, the charge on boron in $B(OH)_3$ is $+2.28$ compared with $+2.43$ in BF_3. The atomic charges vary only slightly between the molecules and ions, just as they hardly change between BF_3 and

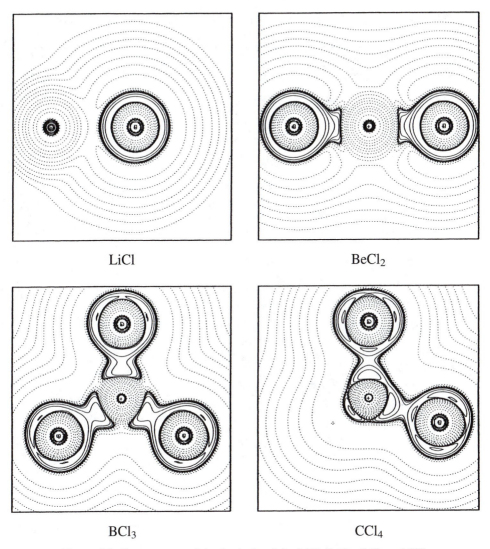

LiCl $BeCl_2$

BCl_3 CCl_4

Figure 8.8 Contour maps of the Laplacian L for LiCl, $BeCl_2$, BCl_3 and CCl_4.

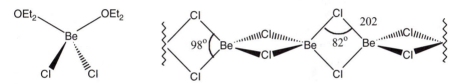

Figure 8.9 Examples of four-coordinated molecules of beryllium: (a) $BeCl_2 \cdot 2Et_2O$ and (b) the polymeric molecule $BeCl_2$.

Table 8.5 Geometrical Parameters, Atomic Charges, and ρ_b Values for Some Be, B, and C Hydroxo Molecules and Ions

Molecule	Bond Lengths (pm)		$\rho_b(au)$	Atomic Charges			
	Calc.	Exp.		$-q(O)$	$q(H)$	$-q(OH)$	$q(A)$
$Be(OH)_2$	142.3		0.133	1.42	0.57	0.85	1.70
$Be(OH)_3^-$	154.6	154.0	0.095	1.37	0.47	0.90	1.69
$Be(OH)_4^{2-}$	168.8		0.065	1.34	0.42	0.93	1.70
$B(OH)_2^+$	125.5		0.267	1.34	0.69	0.65	2.31
$B(OH)_3$	136.9	136.3	0.204	1.32	0.56	0.76	2.28
$B(OH)_4^-$	148.3	147.7	0.153	1.30	0.48	0.82	2.28
$C(OH)_3^+$	128.1		0.358	1.05	0.64	0.41	2.23
$C(OH)_4$	139.3	139.6[a]	0.289	1.04	0.54	0.50	1.99

[a] $C(OMe)_4^-$.

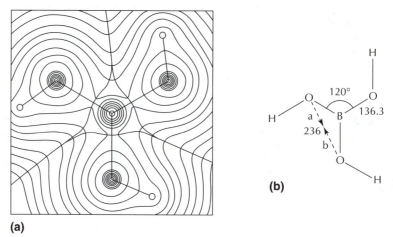

(a)

(b)

Figure 8.10 (a) Contour map of the electron density distribution of $B(OH)_3$. (b) Geometry of $B(OH)_3$.

BF_4^-. Thus the bond length increase from the $A(OH)_2$ to the $A(OH)_4$ molecules must be due primarily to the increasing coordination number.

Table 8.6 gives the bond lengths and bond angles for the trihydroxides and the tetrahydroxides as well as for $B(OMe)_3$, $C(OMe)_4$, and $C(OPh)_4$. All the $A(OX)_3$ molecules have the expected equilateral triangular arrangement of the oxygen atoms, with OAO bond angles of 120°. The OX groups all have the same coplanar orientation, so that in each case the overall symmetry is C_{3h}. The tetrahydroxides as well as $C(OMe)_4$ and $C(OPh)_4$ all have equal AO bond lengths, but they do not have a tetrahedral geometry because none of them has six tetrahedral bond angles. In each case there is a set of two equal angles and a set of four different but equal angles. In $B(OH)_4^-$ there are four angles of 111.1° and two of 106.2°, giv-

Table 8.6 Geometry of A(OX)$_3$ and A(OX)$_4$ Molecules

Molecule	Bond Length (pm)	Bond Angles (°)		Symmetry
Be(OH)$_3^{-a}$	154.0	120		C_{3h}
Be(OH)$_4^{2-}$ ^{a}	168.8	107.8 × 2	110.3 × 4	D_{2d}
B(OH)$_3$	136.3	120		C_{3h}
B(OMe)$_3$	136.8	120		C_{3h}
B(OH)$_4^-$	147.7	106.2 × 2	111.1 × 4	D_{2d}
C(OH)$_3^+$ ^{a}	128.1	120		C_{3h}
C(OH)$_4^a$	139.3	104.3 × 2	112.1 × 4	D_{2d}
C(OH)$_4^a$	139.3	107.2 × 4	114.2 × 2	S_4
C(OMe)$_4$	139.6	106.9 × 4	114.6 × 2	S_4
C(OPh)$_4$	139.4	101.2 × 4	113.8 × 4	D_{2d}

^{a}Calculated data.

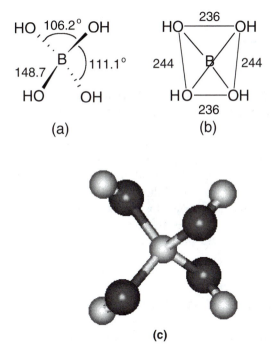

(a)

(b)

(c)

Figure 8.11 Geometry of the B(OH)$_4^-$ ion: (a) bond lengths and bond angles, (b) interligand distances, and (c) ball-and-stick model showing the D_{2d} symmetry.

ing the two different interligand distances of 244 and 236 pm (Figure 8.11). The shorter of these distances is the same as in B(OH)$_3$.

The deviation from a true tetrahedral geometry that we find for the molecules Be(OX)$_4^{2-}$, B(OX)$_4^-$, and C(OX)$_4$ is common to all A(OX)$_4$, A(NX$_2$)$_4$, and A(CX$_2$Y)$_4$ molecules, all of which have two bond angles smaller than, and four greater than, 109.5° or two angles larger than 109.5° and four smaller than 109.5°. In each case the overall symmetry of the molecule, which depends on the relative orientation of the ligands, is D_{2d} or S_4. Some examples

Table 8.7 Symmetries and Bond Angles in Some A(XY)$_4$ Molecules with Distorted AX$_4$ Tetrahedral Structures

Molecule	Symmetry	Bond Angles (°)	
		XAX	XAX
Be(OH)$_4$$^{2-}$ [a]	D_{2d}	107.8 × 2	110.3 × 4
B(OH)$_4$$^-$ [a]	D_{2d}	106.2 × 2	111.1 × 4
B(OMe)$_4$$^-$	D_{2d}	101.7 × 2	113.5 × 4
B(OSO$_2$Cl)$_4$$^-$	S_4	107.4 × 4	113.8 × 2
C(OH)$_4$[a]	D_{2d}	104.3 × 2	112.1 × 4
C(OH)$_4$[a]	S_4	107.2 × 4	114.2 × 2
C(OMe)$_4$	S_4	106.9 × 4	114.6 × 2
C(OC$_6$H$_5$)$_4$	D_{2d}	101.2 × 2	113.8 × 4
C(OC$_6$H$_3$Me$_2$−3,5)$_4$	D_{2d}	100.9 × 2	114. 0 × 4
C(SC$_6$H$_5$)$_4$	S_4	106.3 × 4	116.0 × 2
C(CH$_2$OH)$_4$	S_4	106.7 × 2	110.9 × 4
C(CH$_2$Cl)$_4$	S_4	106.1 × 2	112.9 × 2
C(CH$_2$Cl)$_4$	D_{2d}	108.3 × 4	111.9 × 2

[a]Calculated data.

are given in Table 8.7. These unequal angles and correspondingly unequal interligand distances are not explained by the VSEPR model but can be explained by the LCP model.

The angles in the A(OX)$_4$ molecules deviate from 109.5° because the density of the O atom is perturbed from axial symmetry by the ligand Y. In particular, the density is greater in the direction between the A-O and O-X bonds than in the opposite direction between the lone pairs, because the bonding density is more spread out and closer to the core than the bonding density. It has been shown that as a consequence of this deviation of the density from cylindrical symmetry, the four ligands cannot all pack as closely as possible around the central atom in such a way that all the interligand distances are equal. As a result, the six interligand angles are not equal to 109.5° but two are smaller and four are larger or *vice versa* (Heard et al., 2000).

♦ **8.6   The Nature of the CO and Other Polar Multiple Bonds**

In Section 8.3 we discussed the information that the analysis of electron densities can give on the nature of polar single bonds. In this section we look at the nature of the C=O and other polar multiple bonds. Experimental geometries and calculated bond critical point densities and atomic charges of molecules of the type XYCO in which the CO bond is described as a double bond in their Lewis structures are given in Table 8.8. Consistent with their formulation as double bonds, we can see from Table 8.8 that the CO bonds in these molecules have lengths that are much shorter than the C—OH and C—OMe bonds in H$_2$CO and (MeO)$_2$CO, which are Lewis single bonds. Moreover, the bond critical point densities ρ_b for the C=O bond are larger than for the C—OH, C—OMe, and C—F bonds, which are single bonds in the Lewis structures. In all cases the charge on the C=O oxygen is larger

Table 8.8 Experimental Geometry and Calculated Properties for Molecules with C=O Double Bonds

Molecule	Bond Lengths (pm)		Bond Angles (°)		ρ_b(au)		Atomic Charges		
	CO	CX	XCO	XCX	CO	CX	q(C)	q(O)	q(X)
H_2CO	120.9				0.431		1.25	−1.24	0.01
F_2CO	117.0	131.7	126.2	107.6	0.467	0.297	2.30	−1.09	−0.61
Cl_2CO	117.6	173.8	124.1	111.8	0.458		1.25	−1.05	−0.10
$(HO)_2CO^a$	120.4	133.9	125.7	108.6	0.427	0.314	2.13	−1.17	−1.05
$(MeO)_2CO$	120.3	134.3	126.5	107.0					
CO_2	116.0	—	—	—	0.464	—	2.60	−1.30	—

^{a}Calculated structure.

than −1.0, indicating that these are very polar double bonds. Because both the atomic charges and the bond critical point density have large values, we expect these bonds to be very strong. The C=O bond has an average bond enthalpy of 706 kJ mol^{-1}, which is much larger than that for a C—O bond (335 kJ mol^{-1}) and even larger than that of the B—F bond (613 kJ mol^{-1}), which is the strongest known single bond. It is also larger than that of the nonpolar C=C bond (619 kJ mol^{-1}), consistent with the expected effect of the polarity of the bond on its strength.

As we discussed in Chapter 5, the short length of the CO bond means that for the ligands to remain close packed, the X ligands in an X_2CO molecule are pushed away from the carbon atom, which necessarily increases the length of the CX bonds and decreases the angle between them. Figure 8.12 compares the bond lengths and bond angles in COF_2 and $COCl_2$ with those in the CF_3^+ and CCl_3^+ molecules. That the ligands in all these molecules are indeed close-packed is confirmed by the almost constant X···X interligand distances, which are close to the sum of the ligand radii. As a consequence, the CF bond in F_2CO (131.7 pm) is longer than the CF bond in CF_3^+ (123.5 pm), while the F···F distance has almost the same value in both molecules. Similarly the CCl bond in $COCl_2$ (173.8 pm) is longer than in CCl_3^+ (165.8 pm), while the Cl···Cl distance is the same (288 pm) (Figure 8.12).

The limitations of conventional Lewis structures are evident if we consider that the polarity of the C=O bond is usually described by means of the two resonance structures **1** and **2.**

Considering that the charge on oxygen is very approximately −1, it would seem that we have to conclude that structure **2** is the best representation of this bond. This structure implies that the C=O bond consists of two very different bonds—a nonpolar covalent bond represented by the bond line and a fully ionic bond represented by the two charges. How-

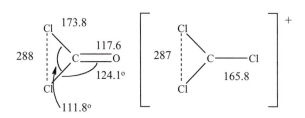

Figure 8.12 The geometry of F_2CO, CF_3^+, Cl_2CO, and CCl_3^+.

ever, this is a purely arbitrary distinction. We can equally well describe the bond as consisting of two partially shared electron pairs, and these two descriptions cannot be distinguished by any experimental method. Moreover, structure **2** does not take into account the size of the actual charge on oxygen, which is somewhat greater than -1. An alternative and somewhat better description of the bond could be given in terms of the two structures **1** and **3**, representing a fully covalent double bond and a fully ionic double bond. This description does not limit the charge on oxygen to -1 and is consistent with the short length and large bond critical point density. In other words, the bond is best considered to be a very polar *double* bond.

8.6.1 The OCF_3^- Ion

The limitations of conventional Lewis structures are further exemplified by the bond lengths in the OCF_3^- molecule (Figure 8.13a), which has been the subject of much discussion because of the difficulty of representing the bonding by means of conventional Lewis structures. The short CO bond length of 122.7 pm is not much longer than the CO bond in methanal (120.9 pm), and this short length has been taken to indicate that the CO bond is a double bond. Representing the CF bonds as single bonds then gives the Lewis structure **4**, which is not usually considered to be an acceptable Lewis structure because the carbon atom is pentavalent and violates the octet rule. Because the CF bonds (139.2 pm) are considerably longer than in CF_4 (131.9 pm), it is generally considered that the bonding is best described in terms of resonance structures such as **5** and three structures like **6**, which obey the octet rule, with **5** being considered to be relatively unimportant because of the short length of the CO bond. It is generally supposed that the length of the CF bonds is accounted for by the no-bond–

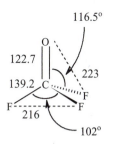

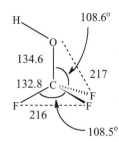

Figure 8.13 The geometry of (a) CF_3O^- and (b) CF_3OH.

single-bond resonance structures for the CF bond. However, these resonance structures only *describe* the bonding in the molecule based on the observed geometry; they do not *explain* the geometry. The explanation of the geometry can be found in the LCP model. As can be seen from the interligand distances in Figure 8.13 the four ligands are close-packed with the F···F and F···O distances close to the sum of the ligand radii. Because the CO bond is so short, the CF bonds are necessarily longer than in CF_4. When an F ligand in CF_4 is replaced by an O ligand, which forms a much shorter bond, the fluorine ligands are pushed away from the carbon, increasing the lengths of the CF bonds and increasing the OCF angle and decreasing the FCF angles so that the O···F and F···F distances can achieve their close-packed values. In the calculated structure of the corresponding alcohol CF_3OH (Figure 8.13b) the CO bond, which is a Lewis single bond, is considerably longer than the CO bond in OCF_3^-, the CF bond is comparable in length to that in CF_4, and the bond angles are close to tetrahedral.

The best Lewis-type representation of the bonding in OCF_3^- would therefore appear to be as in **4**, even though the carbon atom does not obey the octet rule. This molecule can be considered to be a hypervalent molecule of carbon just like the hypervalent molecules of the period 3 elements, such as SF_6. We introduced the atom hypervalent in Chapter 2 and we discuss it in more detail in Chapter 9. But it is important to emphasize that the bonds are very polar. In short, OCF_3^- has one very polar CO double bond and three very polar CF single bonds. A serious limitation of Lewis structures is that they do not give any indication of the polarity of the bonds, and much of the discussion about the nature of the bonding in this molecule has resulted from a lack of appreciation of this limitation.

8.6.2 The Carbonate Ion CO_3^{2-}

The bonding in the carbonate ion is usually represented by the three resonance structures **7–9**

$$O=C\overset{\displaystyle O^-}{\underset{\displaystyle O^-}{\Big\backslash}} \qquad ^-O-C\overset{\displaystyle O}{\underset{\displaystyle O^-}{\Big\backslash}} \qquad ^-O-C\overset{\displaystyle O^-}{\underset{\displaystyle O}{\Big\backslash}}$$

$$(7) \qquad\qquad (8) \qquad\qquad (9)$$

giving a bond order of 1.3, which is considered to be consistent with the observed bond length of 129.4 pm, which in turn is intermediate between the length of a double bond such as that in H_2CO (120.9 pm) and that of a single bond such as that in CH_3OH (141.6 pm). However, this takes no account of the effect on the bond length of the coordination number or of the bond polarity.

It is important to be aware that statements such as "The CO bonds in CO_3^{2-} are short because there is resonance between structures **7, 8,** and **9**" are incorrect because resonance is a not a phenomenon but a description of the bonding in terms of hypothetical Lewis structures.

Table 8.9 lists molecules with CO bonds that in Lewis structures are described as either single or double, together with their lengths, bond critical point densities, and atomic charges. We see that in general Lewis single bonds are longer than double bonds, which in turn are longer than triple bonds. But the range of the lengths of single bonds overlaps those of order 1.3 and 1.5 and comes very close to the range of the double bond lengths.

Several factors appear to affect these bond lengths:

Coordination number of carbon. Molecules in which the carbon is four-coordinated generally have longer CO bonds than those in which carbon is three-coordinated, which in turn are longer than the bonds in CO_2 and CO, consistent with the LCP model.

Table 8.9 Properties of Some CO Bonds

Molecule	B.O.[a]	Bond Lengths (pm)		$\rho_b(au)$	Atomic Charges			Coord. No.	
		Calc.	Exp.		q(C)	q(O)	q(OH)	C	O
HOCO₂⁻	1	145.4	134.6	0. 241	+2.05	−1.24	−0.55	3	2
CH_3OH	1	140.0		0.287	+0.74	−1.24	−0.64	4	2
CH_3OCH_3	1	139.0	141.6	0.273	+0.78	−1.29		4	2
$C(OH)_4$	1	139.3		0.289	+1.98	−1.04	−0.50	4	2
FH_2COH	1	136.1		0.290	+1.35	−1.26	−0.64	4	2
(HO)₂CO	1	133.9	134.3	0.314	+2.13	−1.05	−0.60	3	2
CH_3O^-	1	132.6		0.332	+1.44	−1.04		4	1
F_3COH	1	132.8		0.337	+2.78	−1.24	−0.62	4	2
CO_3^{2-}	1.3	130.8	129.4	0.339	+2.10	−1.34		3	1
FH_2CO^-	1	125.8		0.383	+1.48	−1.45		4	1
HOCO₂⁻	1.5	125.1		0.385	+2.05	−1.24		3	1
F_3CO^-	1	122.7		0.437	+2.16	−1.26		4	1
(HO)₂CO	2	120.4	120.3	0.427	+2.13	−1.17	−0.60	3	1
H_2CO	2	118.3	120.9	0.431	+1.25	−1.24		3	1
Cl_2CO	2	117.2	117.6	0.458	+1.25	−1.05		3	1
F_2CO	2	117.1	117.2	0.467	+2.30	−1.29		3	1
CO_2	2	114.3	116.0	0.483	+2.59	−1.30		2	1
CO	3	111.4	112.8	0.510	+1.35	−1.35		1	1

[a]Bond order as given by the conventional Lewis structure. Italic indicates which of the two bonds the data is given for.

Coordination number of oxygen. The CO bonds in molecules in which the oxygen is two-coordinated as in OX ligands, which are described as Lewis single bonds, are generally longer than those in which the oxygen is one-coordinated (i.e., is in a terminal position), and are described as C=O bonds, but there is no clear division between the two groups.

Atomic charges. Although the atomic charges on carbon and oxygen in a CO bond are large their effect on the bond length is often obscured by the other factors and is clear only when closely related molecules are compared. For example, in the series F_3CO^-, FH_2CO^-, and H_3CO^- (Table 8.9) the increase in the CO bond length from 122.7 to 125.8 to 132.6 pm correlates with decrease in the product of the C and O atomic charges from 2.72 to 2.15 to 1.50. Similarly, in the series F_3COH, FH_2COH, H_3COH, the increase in the CO bond length from 132.8 to 136.1 to 140.0 correlates with the product of the C and O atomic charges, which decreases from 3.45 to 1.70 to 0.50.

It is clear that the Lewis structures of these molecules are only very crude approximations and that there is no typical single C—O or double C=O bond length. The data in Table 8.9 show that the concept of bond order, based on two or more resonance structures, such as the bond order of 1.3 in the carbonate ion, although a qualitatively useful idea, can have no quantitative significance. In general, any attempt to explain the length of a bond in terms of resonance structures is, at best, only a very rough approximation. Such structures assume that electrons are either bonding or nonbonding, but the total electron density is a continuous function in which different electrons cannot be distinguished. Therefore in principle all the oxygen electrons may be involved to a greater or a lesser extent in a CO bond. The effective number of electrons that may be considered to be bonding electrons is, in general, not just two, four, or six but can vary continuously. That nonbonding electrons cannot in general be clearly distinguished from bonding electrons is also shown by the function L. As we saw in Chapter 7 the charge concentrations attributed to the nonbonding and bonding electrons in a molecule such as SCl_2 (Figure 7.3) are only maxima in a continuous charge concentration.

We have seen in Chapter 6 (Figure 6.16) that there is only one bond path between carbon and oxygen in H_2CO and this is the case for any CO bond, although this bond may be due to one, two, or three or any intermediate number of electron pairs. In this respect a CO bond is just like a CC bond.

A striking feature of the data in Table 8.9 is that the bond critical point density ρ_b is a continuous and almost linear function of the bond length, as can be seen in Figure 8.14. Does the bond length determine ρ_b or does ρ_b determine the bond length? As we have seen, bond lengths appear to be determined by several factors, such as coordination number and atomic charges, so it might appear that the bond critical point density is determined by the bond length. However, the relationships between the various atomic and bond properties need further investigation, and the question just posed is in general not easily answered

The foregoing considerations concerning CO bonds must clearly apply to all other bonds to oxygen, which in a Lewis structure are described as either single, double, or triple (including, e.g., BO, NO, PO, SO and ClO bonds). They also apply to bonds to nitrogen, which similarly are described as single, double, or triple.

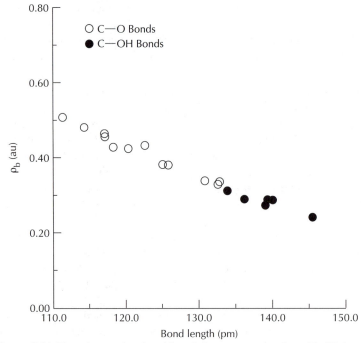

Figure 8.14 Plot of ρ_b against bond length for some molecules with CO bonds.

8.6.3. The Carbon Monoxide Molecule CO

The nature of the bonding in this molecule has been the cause of considerable discussion. Its short length (112.8 pm) and its great strength (bond dissociation enthalpy 1072 kJ mol^{-1}) are consistent with the usual triple-bond Lewis structure

$$:C^-:::O:^+$$

This structure has large formal charges on the atoms which are, however, opposite in sign to the calculated charges of $+1.35$ and -1.35 (Table 8.9). Moreover, these large charges suggest that the molecule should have a large dipole moment. However, large atomic dipoles of 14.42×10^{-30} C·m on the carbon atom and 9.00×10^{-30} C·m on the oxygen atom can be calculated from the electron density distribution. They arise from the polarization of the charge density of each of the atoms, as can be seen in the contour maps of the electron density and its Laplacian shown in Figure 8.15. These atomic dipole moments oppose the dipole moment of 24.39 all arising from the large atomic charges all we have discussed in Chapters 2 and 6, giving an overall very small dipole moment of only 0.97×10^{-30} C·m, in good agreement with the measured value of 0.37×10^{-30} C·m.

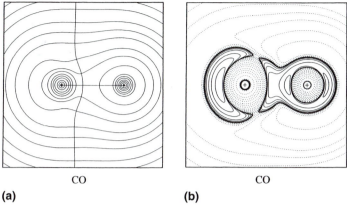

	CO		CO
	(a)		**(b)**

Figure 8.15 Contour maps for the CO molecule: (a) the electron density distribution ρ and (b) the Laplacian L.

◆ 8.7 Bonding and Geometry of the Molecules of Nitrogen

Nitrogen resembles the other elements in group 15, such as phosphorus. However, an important difference between nitrogen and phosphorus is that five- and six-coordinated molecules of nitrogen in its +V oxidation state such as NF_5 and NF_6^- are not known, principally because the nitrogen atom is too small to allow five or six fluorine atoms to be packed closely around it. In contrast, there are many molecules of P(+V) with a coordination number of five or six such as PF_5, PCl_5, PF_6^-, and PCl_6^-. In molecules in which nitrogen is in the +III oxidation state it uses only three of its electrons to form bonds, leaving an unshared or lone pair in its valence shell. We first consider molecules in which nitrogen is in the +III oxidation state.

8.7.1 N(+III) Molecules

According to the VSEPR model, NX_3E molecules are expected to have a triangular pyramidal geometry. Experimental data for some NX_3E molecules are given in Table 8.10 and Figure

Table 8.10 Bond Angles, Bond Lengths, and Interligand Distances in NX_3E Molecules

Molecule	N—X (pm)	XNX (°)	X···X (pm) Observed	X···X (pm) Predicted
NH_3	99.7	107.2	161	164
NF_3	136.5	102.3	212	214
NCl_3	175	106.8	280	284
$N(CH_3)_3$	145.8	110.9	240	238
$N(CF_3)_3$	142.6	117.9	244	238
$N(SiH_3)_3$	173.4	120.0	300	
$N(SCF_3)_3$	170.5	118.8	294	

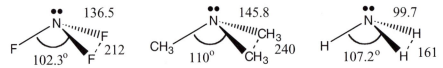

Figure 8.16 The geometry of some trigonal pyramidal molecules of nitrogen.

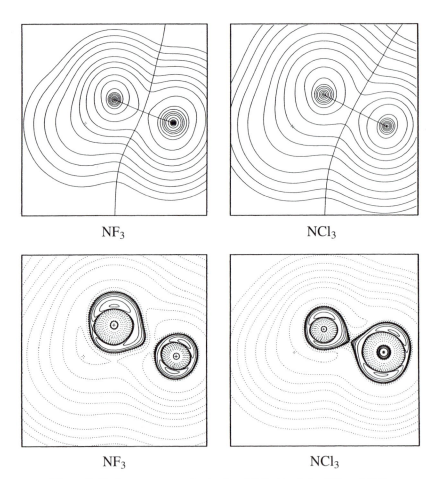

NF₃

NCl₃

NF₃

NCl₃

Figure 8.17 Contour maps of ρ (top) and L (bottom) for NF_3 and NCl_3.

8.16. Only the molecules NH_3, NF_3, and NCl_3 have bond angles that are smaller than tetrahedral, while that in $N(CH_3)_3$ is very slightly larger than tetrahedral. The bond angles in these molecules are close to or less than tetrahedral because in these molecules the electronegativity of the ligand is greater than, or not much smaller than, that of nitrogen, so that the lone pair is well localized and behaves like a pseudoligand. Contour maps of the electron density and L are given in Figure 8.17 for NF_3 and NCl_3. In the maps of L there are in each case three bonding CCs as well as a nonbonding (lone pair) CC in the valence shell of nitrogen, making a total of four CCs in an approximately tetrahedral arrangement. These charge concentrations result from the partial localization of the valence shell electrons into four pairs, consistent with the VSEPR model. Evidence for a nonbonding pair is also seen in the electron density maps, which show a distinct bulge in the electron density of the nitrogen atom in the lone pair direction. In these molecules in which there is a well-localized lone pair, the ligands are pushed together untill the ligand–ligand distance is twice the ligand radius.

The other molecules in Table 8.10, have bond angles that are larger than tetrahedral. In $N(SiH_3)_3$ the electronegativity of nitrogen is considerably higher than that of silicon, so the valence shell electrons of nitrogen are not strongly localized into bonding and nonbonding pairs, as shown by the absence of well-defined nonbonding CC in the Laplacian map in Figure 8.18. As a consequence, ligand–ligand repulsions dominate the geometry, and the bond angle attains the maximum value of 120°. From the absence of any charge concentrations in

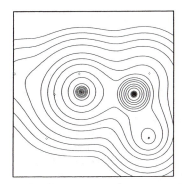

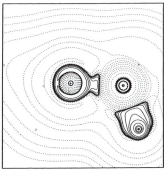

$N(SiH_3)_3$

Figure 8.18 Contour maps of ρ (top) and L (bottom) for $N(SiH_3)$ in a plane through N, Si, and H.

the valence shell of silicon, we see also that much of the valence shell electron density has been transferred from the silicon atom to the much more electronegative nitrogen atom. The reasons for the large bond angles in N(CF$_3$)$_3$ and N(SCF$_3$) are not clear.

Table 8.11 gives the bond lengths and angles in some NX$_2$YE molecules. The bond angles and lengths are determined primarily by the close packing of the ligands and the lone pair domain around the central nitrogen, as shown by the generally good agreement between the observed interligand distances and those predicted from the ligand radii.

Molecules of the type ONXE where the NO bond is a Lewis double bond are expected to be angular. Some molecules of this type are illustrated in Figure 8.19. In each case the bond angle is less than 120°, consistent with the presence of the lone pair. The bond angle in the nitrosyl halides decreases with decreasing size of the halogen as expected, and the O$\cdots$Hal distance is equal to the sum of the radii for ligands bonded to nitrogen. The bond length of 126 pm in the nitrite ion NO$_2^-$ is intermediate between that of the Lewis double bonds in the nitrosyl halides and the Lewis NO single bond in nitrous acid, H—O—N=O, consistent with the intermediate nature of this bond as expressed by the two resonance structures. The NO$_2$ molecule in which there is a single nonbonding electron on nitrogen, and which is an example of an NX$_2$e molecule, where e represents a single unpaired electron, has a bond angle (134°); this larger than 120°, consistent with a single nonbonding electron having a smaller domain than a lone pair. The large O$\cdots$O distance of 224 pm in this molecule suggests that the oxygen ligands are not close-packed because of the small repulsive effect of the single nonbonding electron. In NO$_2^+$ in which nitrogen is in the +V state there are no nonbonding electrons, so the molecule is linear.

8.7.2 N (+V) Molecules

Although there are a relatively large number of N(+V) molecules, the absence of any with a coordination number greater than four must be attributed to the small size of the nitrogen

Table 8.11 Bond Lengths, Bond Angles, and Interligand Distances in NX$_2$YE Molecules

Molecule	Bond	Length (pm)	Angle	Angle (°)	Ligand X$\cdots$X	Interligand Distances (pm) Observed	Interligand Distances (pm) Predicted
NH$_2$CH$_3$	N—H	101.1	HNH	105.8	H$\cdots$H	161	164
	N—C	147.7	HNC	112.1	H$\cdots$C	208	202
NH(CH$_3$)$_2$	N—H	102.2	HNC	108.8	H$\cdots$C	204	202
	N—C	146.6	CNC	111.6	C$\cdots$C	243	240
NH$_2$F	N—H	102.9	HNH	101.6	H$\cdots$H	161	164
	N—F	140.0	HNF	99.8	H$\cdots$F	191	189
NF(CH$_3$)$_2$	N—F	144.7	CNC	112.0	C$\cdots$C	242	240
	N—C	146.2	FNC	104.6	C$\cdots$F	229	227
NF$_2$CH$_3$	N—F	141.3	FNF	101.0	F$\cdots$F	218	214
	N—C	144.9	FNC	103.6	C$\cdots$F	229	227
NF$_2$Cl	N—F	138.2	FNF	103	F$\cdots$F	216	214
	N—Cl	173.0	FNCl	105	F$\cdots$Cl	248	247
NCl$_2$CH$_3$	N—Cl	174	ClNCl	108	Cl$\cdots$Cl	282	280
	N—C	142	ClNC	109	Cl$\cdots$C	262	260

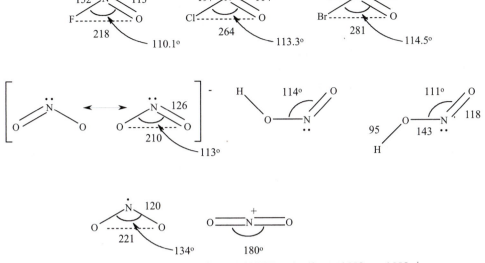

Figure 8.19 The geometry of some ONXE molecules and NO_2^- and NO_2^+.

Table 8.12 Bond Lengths, Bond Angles, and Interligand Distances in Some NX_4^+ and ONX_3 Molecules

Molecule	Bond Lengths (pm)		Bond Angles (°)		Interligand Distances (pm)	
	N—X	N—O	XNX	XNO	X···X	X···O
NH_4^+	103.2		109.5		161	
NF_4^+	130		109.5		212	
$N(CH_3)_4^+$	151		109.5		247	
ONF_3	143.1	115.8	100.8	118.1	216	213
$ON(CH_3)_3$	147.7	138.8	109.0	109.9	240	
ONH_3^a	103.0	136.8	106.2	112.6	165	

[a]Calculated data.

atom. NX_4^+ molecules have the expected tetrahedral geometry. Bond lengths and bond angles for molecules of this type are given in Table 8.12. When the ligands are all the same, they have a regular tetrahedral geometry as in NF_4^+ and $N(CH_3)_4^+$. The NF bonds in NF_4^+ (130 pm) are shorter than in NF_3 (136.5 pm) because of the absence of the repulsive effect of the lone pair. The lone pair occupies a larger domain than the bonding pairs, decreasing the FNF bond angle and correspondingly increasing the bond length with the ligands remaining in contact, as shown by the F···F distance, which is the same (212 pm) as in NF_4^+.

Molecules of the type ONX_3 are of particular interest because the NO bond in ONF_3 is very short (115.8 pm) (Figure 8.20), as we have seen is also the case for the CO bond in the isoelectronic molecule OCF_3^- (122.7 pm). Hence the NF bonds are considerably longer than in NF_4^+ and there are corresponding considerable deviations in the bond angles from the tetrahedral value, with the XNO angles being larger than the XNX angles. In $ON(CH_3)_3$ and in the calculated structure of ONH_3, in contrast, the NO bond is much longer (138.8 and 136.8 pm, respectively), and the NC and NH bonds are much closer to the lengths of the corresponding bonds in $N(CH_3)_4^+$ (151 pm) and NH_4^+ (103.2 pm) than in ONF_3. Because of its short length, the NO bond in ONF_3 it is best considered to be a double bond like the CO bond in COF_3^-, which we discussed in Section 8.6.1.

There are many molecules of the type XNO_2, and all have the expected triangular planar geometry. They include the nitrate ion, nitric acid, nitryl fluoride, and several oxides of nitrogen. Bond lengths, bond angles, and ligand–ligand distances are given in Figure 8.21 and Table 8.13. In the nitrate ion and XNO_2 molecules the O···O distances are close to the average value of 218 pm and are consistent with the close packing of the O ligands around the nitrogen with a ligand radius of 109 pm.

The representation of the bonding in the oxo molecules of nitrogen by means of Lewis structures encounters the same problems we discussed at some length for CO bonds in Section 8.6. NO bond lengths vary greatly from, for example, 141 pm for the HO–N bond in HNO_3, which is formally a single bond to 115 pm in NO_2^+, which has formal double bonds, and to 106 pm in NO^+, in which the bond is formally a triple bond. The NO bonds in XNO_2 molecules (Table 8.13) and in ONF_3 have lengths in the range of 118–122 pm and so are much closer in length to a double bond than a single bond, suggesting that the bonds would

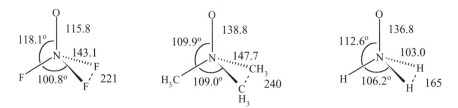

Figure 8.20 The geometry of ONF_3, $ON(CH_3)_3$, and ONH_3.

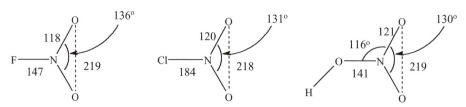

Figure 8.21 The geometry of some XNO_2 molecules.

Table 8.13 Bond Angles, Bond Lengths, and Interligand Distances in the NO_2 Group in XNO_2 Molecules

Molecule	NO (pm)	ONO (°)	O···O (pm)
NO_3^-	125	120	216
HNO_3	121,120	130	218
$ONNO_2$	122,120	130	219
O_2NNO_2	119	136	220
O_2NONO_2	118	133	217
FNO_2	118	136	217
$ClNO_2$	120	131	218

be better described as double bonds **10** rather than in the customary manner with resonance structures based on the octet rule **11**, provided we recognize that the double bonds in **10** are highly polar.

NX$_2$ molecules have the expected linear geometry as shown in Figure 8.22 in which the molecules are represented by their Lewis structures with double bonds. The short lengths of the bonds are consistent with this very approximate picture, but the NN bonds vary in their nature from molecule to molecule, as do CO and NO bonds. Although the azide ion is linear with equal bond lengths, in hydrogen azide and chlorine azide the two NN bonds have different lengths, and azide group is slightly bent at the central nitrogen. The different bond lengths are more consistent with the structure in which there is a double and a triple NN bond (**12**) than the structure implied by the Lewis model, in which the central nitrogen has an octet (**13**). No explanation has been given for this unexpected bond angle at the central nitrogen, but it may well be due to an unsymmetrical electron density around the central nitrogen.

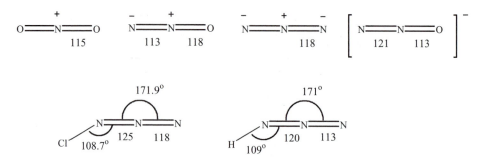

Figure 8.22 The geometry of some NX$_2$ molecules.

◆ 8.8 The Geometry of the Molecules of Oxygen

The possible geometries for oxygen are OX$_2$E$_2$ angular, OX$_3$E trigonal pyramidal (in which oxygen has a formal positive charge), OX$_4$ tetrahedral (in which oxygen has a formal double positive charge), and OX$_2$E angular (in which one of the ligands is doubly bonded in the Lewis structure and oxygen has a formal single positive charge).

8.8.1 OX$_2$E$_2$ Angular Geometry

According to the VSEPR model, bond angles in OX$_2$ molecules should be smaller than the tetrahedral angle. However, we can see in Table 8.14 that only H$_2$O and F$_2$O have bond angles that are less than the tetrahedral value, while Cl$_2$O and (CH$_3$)$_2$O have bond angles that are slightly larger than tetrahedral. The remaining molecules in the table have much larger bond angles. We noted very similar trends in the bond angles in NX$_3$E molecules. Just as in N(SiH$_3$)$_3$, the ligand atoms are considerably less electronegative than the central atom so the valence shell electrons of oxygen are not well localized into pairs. As a consequence, the bond angles in these molecules are determined primarily by ligand-ligand repulsions.

The large bond angle of 144° in (H$_3$Si)$_2$O has been the subject of much discussion. Not only is the bond angle large, but the bonds are short if they are compared to the sum of the covalent radii, which is 182 pm. This apparently short bond length has often been interpreted as indicating double-bond character resulting from the delocalization of the oxygen lone pair electrons into vacant d orbitals on silicon, which can be represented by resonance structures such as **14** or into Si—X antibonding orbitals, which can be approximately represented by resonance structures such as **15**.

(14)   (15)

Table 8.14 Bond Lengths, Bond Angles, and
Interligand Distances in OX_2E_2 Molecules

Molecule	OX (pm)	XOX (°)	X⋯X (pm)
H_2O	95.8	104.5	152
F_2O	140.9	103.3	220
Cl_2O	170.0	110.9	280
$(CH_3)_2O$	141.0	111.7	234
$(SiH_3)_2O$	163.4	144.1	311
$(GeH_3)_2O$	176.6	126.5	315
$(Cl_3Si)_2O$	159.2	146	304
$(F_3Si)_2O$	158	156	309
$(Me_3Si)_2O$	163	148	313
$(F_2P)_2O$	153.3	135.2	292
$[(CH_3)_2B]_2O$	135.9	144	258

Table 8.15 Bond Lengths, Bond Angles, Bond Critical Point Densities, and Atomic Charges for $(H_nX)_2O$ Molecules

Molecule	Bond Lengths (pm)[a]		Bond Angle[a]	ρ_b (au)	Atomic Charges		
	X—O	X—H	XOX (°)		q(O)	q(X)	q(H)
Li_2O	159.6 (160)	—	180.0 (180)	0.080	−1.82	+0.91	—
$(HBe)_2O$	139.6	133.2	180.0	0.148	−1.79	+1.74	−0.85
$(H_2B)_2O$	135.4	119.0	126.9	0.209	−1.68	+2.27	−0.72
$(H_3C)_2O$	139.0 (141.0)	108.4 (109.6)	113.9 (111.7)	0.273	−1.29	+0.78	−0.05
$(H_2N)_2O$	138.9	99.0	109.8	0.329	−0.51	−0.46	+0.36
$(HO)_2O$	136.5	94.5	107.8	0.357	−0 .04	−0.62	+0.64
F_2O	133.6 (140.5)	—	103.5 (103.1)	0.370	+0.33	−0.17	—
Na_2O	197.3	—	180.0	0.053	−1.77	+0.88	—
$(HMg)_2O$	178.2	170.5	180.0	0.082	−1.77	+1.69	−0.81
$(H_2Al)_2O$	167.1	157.5	180.0	0.11 3	−1.76	+2.45	−0.78
$(H_3Si)_2O$	162.1 (163.4)	147.2 (148.6)	148.3 (144.1)	0.141	−1.72	+3.05	−0.73
$(H_2P)_2O$	163.6	140.9	129.8	0.161	−1.59	+1.97	−0.59
$(HS)_2O$	165.4	132.8	119.1	0.190	−1.24	+0.75	−0.13
Cl_2O	166.5 (169.6)	—	112.8 (111.2)	0.214	−0.66	+0.33	—

[a]Experimentally observed values are given in parentheses.

However, as we have seen, the comparison of the bond lengths with sum of the covalent radii is not a valid procedure because bonding radii are dependent on the atomic charges, which are large in this molecule.

The bond angles in these OX_2 molecules can be better understood in the light of the information obtained from the results of the analysis of the calculated electron density distribution given in Table 8.15 for the series of molecules Li_2O, $(BeH)_2O$, $(BH_2)_2$ O, $(CH_3)_2O$, $(NH_2)_2O$, $(HO)_2O$, F_2O, and the corresponding period 3 molecules. Most of these molecules are unknown; for the known molecules F_2O, Cl_2O, $(CH_3)_2O$, $(SiH_3)_2O$, and Li_2O, however, there is good agreement between the calculated and experimental geometrical parameters. The importance of the calculated properties of the unknown molecules is that they show that molecules such as Li_2O and $[(CH_3)_3Si]_2O$ are not anomalous but follow the trends exhibited in both these series. The bond angle decreases in each series from 180° to a value approximately equal to or less than the tetrahedral angle in both series and correlates well with the electronegativity of the ligand. With increasing electronegativity of the ligands, the bonds become more covalent and the valence shell electrons of oxygen become more localized. The charges on the O and A atoms decrease from large values for the first members of each series to rather small values for the later members consistent with the expected trend from predominately ionic molecules to predominately covalent molecules. It is the strong attraction between the large charges on the O and Si atoms in $(H_3Si)_2O$ that causes these bonds to be very short and to be correspondingly strong as shown by the mean bond enthalpy of the Si—O bond of 464 kJ mol^{-1} (Table 2.8). We have noted before, in the case of BF_3 for example, the great strength and short lengths of very polar bonds in which the atomic charges are large.

Contour plots of L, the Laplacian of the electron density for $(CH_3)_2O$ and $(SiH_3)_2O$, are given in Figure 8.23. For $(CH_3)_2O$ we see four well-developed charge concentrations, but in $(SiH_3)_2O$ the valence shell charge concentration of the oxygen shows two very weak bonding charge concentrations and what appears to be a very weak nonbonding charge concentration. This latter feature is, however, just the cross section of a very weak nonbonding CC that is spread out almost completely around the silicon atom in an almost complete torus. These features of the Laplacian map show that the oxygen valence shell electrons are only very poorly localized into pairs and therefore have only a small influence on the geometry, which is dominated by the repulsions between the SiH_3 ligands. In the linear Li_2O molecule, which is predominately ionic, and in which the central oxygen closely resembles an O^{2-}, there is essentially no localization of the valence shell electrons on oxygen into pairs so that the linear geometry is due entirely to ligand–ligand repulsion.

The bond angle in digermoxane $(H_3Ge)_2O$ is smaller and the charge on oxygen smaller than in $(H_3Si)_2O$, consistent with the greater electronegativity of germanium (2.0) than of silicon (1.7). Moreover, the bond angle and charge on oxygen in $(GeH_3)_2O$ are close to those in $(H_2P)_2O$, which is consistent with the electronegativity of phosphorus (2.1), which is nearly the same as that of germanium (Table 1.2).

We saw in Section 8.7 that the bond angle at nitrogen increases in a similar way with decreasing electronegativity of the ligand in the series NF_3 (102.3°), NH_3 (107.2°), $N(CH_3)_3$ (110.9°), and $N(SiH_3)_3$ (120°). NX_3 molecules more easily achieve a planar geometry than OX_2 molecules achieve a linear geometry because the number of ligand–ligand repulsions is greater in NX_3 molecules than in OX_2 and the corresponding increase in the bond angle is smaller for NX_3 molecules.

Figure 8.24 gives the bond lengths and angles for some HOX molecules. At first sight it might seem surprising that the bond angles in these HOX molecules are smaller than in both H_2O and X_2O. But we can see that the $X \cdots H$ interligand distances are close to the average of $X \cdots X$ distances in the X_2O molecules and the $H \cdots H$ distance in the H_2O molecule consistent with the LCP model. The small angles in the XOH molecules are simply a consequence of the different lengths of the HO and XO bonds.

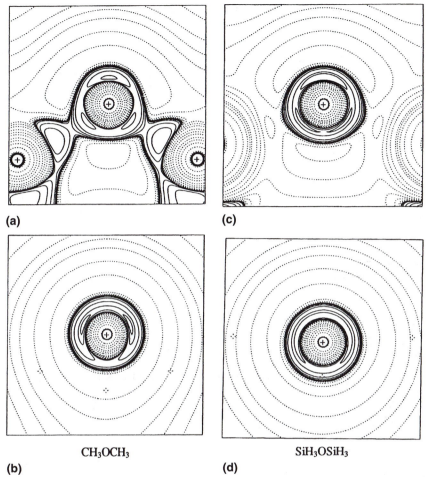

(a)

(c)

CH₃OCH₃

SiH₃OSiH₃

(b)

(d)

Figure 8.23 Contour plots of L for the oxygen atom in $(CH_3)_2O$ (a) in the molecular plane and (b) perpendicular to the molecular plane, and for the oxygen atom in $(SiH_3)_2O$ (c) in the molecular plane and (d) perpendicular to the molecular plane through the oxygen.

8.8.2 OX₃E Trigonal Pyramidal Geometry

The hydronium ion H_3O^+ has the trigonal pyramidal geometry, but the bond angles vary over a wide range depending on the accompanying anion with which the H_3O^+ ion is hydrogen-bonded. Ethers such as dimethyl ether combine with many acceptor molecule such as BF_3 to form adducts in which oxygen is three-coordinated and is therefore expected to have the AX_3E trigonal pyramidal geometry as, for example, in $BF_3 \cdot OMe_2$ (Figure 8.25).

8.8.3 OX₄ Tetrahedral Geometry

An unusual geometry for oxygen in discrete molecules, the OX_4 tetrahedral geometry is found in the oxotetracarboxylates of beryllium such as oxohexaacetatotetraberyllium Be_4O

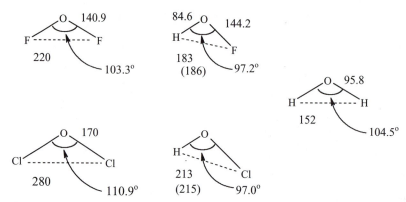

Figure 8.24 The geometry of the molecules F_2O, HOF, and H_2O, Cl_2O, and HOCl

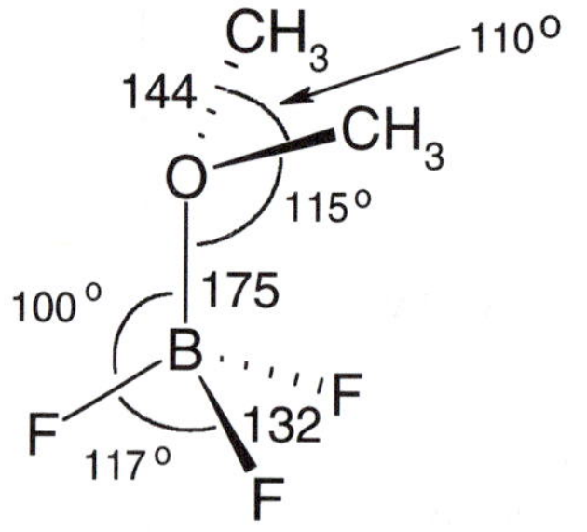

Figure 8.25 The trigonal pyramidal OX_3E molecule $BF_3 \cdot O(CH_3)_2$.

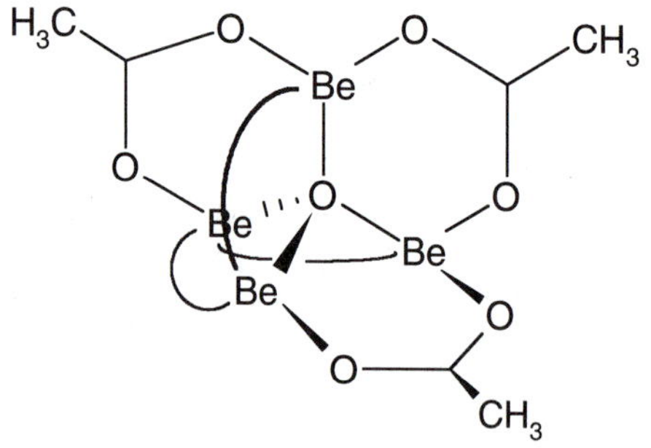

Figure 8.26 The four-coordinated oxygen atom in basic beryllium acetate $OBe_4(CH_3CO_2)_6$. Three of the acetate groups are shown as curved lines.

$(CH_3COO)_6$ (Figure 8.26). Although this geometry is independent of the nature of the BeO bonds, they are expected to be predominately ionic in this molecule. The same geometry is also found in many three-dimensional infinite structures such beryllium oxide.

◆ 8.9   The Geometry of the Molecules of Fluorine

Fluorine normally forms only one bond and so is usually a terminal ligand. Nevertheless fluorine often acts as a bridging atom, particularly in fluorides, for example, in $Sb_2F_{11}^-$ (Fig-

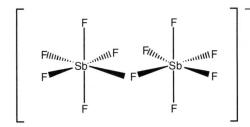

Figure 8.27 The $Sb_2F_{11}^-$ ion in which there is a bridging fluorine linking two SbF_5 groups, each antimony having an approximately octahedral geometry.

ure 8.27). The bond angle at fluorine in the $Sb_2F_{11}^-$ ion in different crystal structures varies from 140° to 180° consistent with the large difference between the electronegativities of fluorine and antimony and the consequent very weak localization of the valence electrons of fluorine into pairs. These bridging bonds are therefore very ionic. They occur frequently in fluorides because the large charges on both atoms lead to a strong, predominately ionic, $F\cdots Sb$ interaction.

▶ Further Reading

For further information on bonding in the molecules of the period 2 elements, the following books are useful.

F. A. Cotton, G. Wilkinson, C. A. Murillo, and M. Bochman, *Advanced Inorganic Chemistry,* 6th ed., 1999, Wiley, New York.
  A comprehensive reference book that gives information on structure and also on the preparation and properties of inorganic substances. Earlier editions are still a useful source of information, particularly on bonding models.
R. J. Gillespie and I. Hargittai, *The VSEPR Model of Molecular Geometry,* 1991, Allyn & Bacon, Boston.
  Chapter 4 describes the VSEPR model for the main group second-period elements.
N. N. Greenwood and A. Earnshaw, *Chemistry of the Elements,* 1984, Pergamon Press, Oxford.
  A comprehensive reference book; particularly strong on the main group elements.
J. E. Huheey, E. A. Keiter, and R. L. Keiter *Inorganic Chemistry,* 4th ed., 1993, HarperCollins, New York.
D. M. P. Mingos, *Essential Trends in Inorganic Chemistry,* 1998, Oxford University Press, Oxford.
D. M. P. Mingos and D. J. Wales, *Introduction to Cluster Chemistry,* 1990, Prentice-Hall, Englewood Cliffs, NJ.
  A useful introduction to the structures of cluster molecules and the electron counting rules proposed by Wade and others.
N. C. Norman, *Periodicity and the p-Block Elements,* 1994, Oxford University Press, Oxford.
D. F. Shriver, P. W. Atkins, and C. H. Langford, *Inorganic Chemistry,* 1990, Freeman, New York.

▶ References

Most of the numerical data quoted in the tables and the text can be found in the foregoing books or in the following articles, where additional references are quoted.

R. J. Gillespie and E. A. Robinson, *Adv. Mol. Struct. Res. 4,* 1, 1998.

R. J. Gillespie, I. Bytheway, and E. A. Robinson, *Inorg. Chem. 37,* 2811, 1998a.

R. J. Gillespie, E. A. Robinson, and G. L. Heard, *Inorg. Chem. 37,* 6884, 1998b.

G. L. Heard, R. J. Gillespie, and D. W. H. Rankin, *J. Mol. Struct. 520,* 237 (2000).

E. A. Robinson, S. A. Johnson, T.-H. Tang, and R. J. Gillespie, *Inorg. Chem. 36,* 3032, 1997.

E. A. Robinson, G. L. Heard, and R. J. Gillespie, *J. Mol. Struct. 485–486,* 305, 1999.

B. D. Rowsell, R. J. Gillespie, and G. L. Heard, *Inorg. Chem. 38,* 4659, 1999.

C H A P T E R

9

MOLECULES OF THE ELEMENTS
OF PERIODS 3–6

■ ■ ■

◆ 9.1 Introduction

As we have seen in earlier chapters, with the exception of the few molecules discussed in Box 8.1 the elements of period 2 have four or fewer ligands. In contrast, there are a large number of molecules of the elements in periods 3 and 4 that have coordination numbers of five and six, and the still larger atoms of the elements of periods 5 and 6 can have coordination numbers as high as seven or eight. For example, sulfur forms the molecule SF_6 and phosphorus the molecule $P(C_6H_5)_5$. This important difference between the geometry of period 2 molecules and those of period 3 and beyond is due very largely to the smaller size of the atoms of the period 2 elements.

The Lewis diagrams of molecules with coordination numbers greater than four imply that there are more than four electron pairs in the valence shell of the central atom and therefore they do not obey the octet rule. Many molecules from period 3 and beyond have lone pairs in the valence shell of the central atom in their Lewis diagram and so may have more than four pairs of electrons in their valence shell even though they have only four or fewer ligands. For example, the molecule $:Te(CH_3)_2Cl_2$ has five electron pairs in the valence shell of Te. We will call the total number of ligands and lone pairs in a valence shell the ligand–lone pair (LLP) coordination number, or LLPCN for short, to avoid confusion with the conventional use of the term coordination number (CN), which denotes the number of ligands only. So for the atoms of the period 2 elements for which LLPCN has a maximum value of four the possible molecular types are AX_2, AX_3, AX_2E, AX_4, AX_3E, and AX_2E_2. The molecules of the elements of periods 3 and 4 can, however, have the following additional geometries that we do not find for the elements of period 2: for LLPCN = 5, AX_5, AX_4E, AX_3E_2, and for LLPCN = 6, AX_6, AX_5E, AX_4E_2. There are no known examples of AX_3E_3 molecules. Molecules with LLPCN ≥ 7 have still other geometries, which we describe in Section 9.7.

Molecules that have more than four electron pairs in the valence shell of the central atom in the Lewis diagram and therefore do not obey the octet rule are often called **hypervalent**

molecules. Because of the importance of this type of molecule for the elements of period 3 and beyond, and because the meaning of the concept of hypervalence is often misunderstood and has been the subject of much controversial discussion, we discuss hypervalent molecules in the following section.

In the later sections we discuss the geometry of some typical molecules of the period 3 elements.

◆ 9.2 Hypervalence

Because they do not obey the octet rule, hypervalent molecules have often been thought to involve some type of bonding that is not found in period 2 molecules. Ideas concerning the nature of this bonding have developed along a somewhat tortuous path that it is interesting and instructive to follow. We will in the end conclude that the nature of the bonding in these molecules is not different in type from that in related period 2 molecules and that there is therefore little justification for the continued use of this concept.

The simplest explanation of the existence of molecules with a LLP coordination number greater than 4 (i.e., hypervalent molecules) is that, as the central atom gets larger from period 2, it is possible to pack an increasing number of ligands around the central atom. This simple explanation has in the past been largely ignored in favor of arguments based on orbital models although it now is becoming increasingly accepted.

The octet rule was proposed by Lewis when he found that, on counting the electrons of a bonding pair as contributing to the valence shell of *both* bonded atoms, in almost all the molecules with which he was familiar, each atom, except hydrogen, had eight electrons in its valence shell. He called this observation the rule of eight, although it later became known as the octet rule. Lewis was aware of a small number of molecules such as PCl_5 and SF_6 that were exceptions to his octet rule, because in their Lewis structures they have ten and twelve electrons respectively in the valence shell of the central atom. But he did not consider these few exceptions to be of any great importance because he regarded the rule of two (electrons in the vast majority of molecules are found in pairs) as fundamentally more important. Since that time many more hypervalent molecules have been discovered, and it is no longer reasonable to consider them as minor, unimportant exceptions.

Lewis considered covalent and ionic bonds to be two extremes of the same general type of bond in which an electron pair is shared between two atoms contributing to the valence shell of both the bonded atoms. In other words, in writing his structures Lewis took no account of the polarity of bonds. As we will see much of the subsequent controversy concerning hypervalent molecules has arisen because of attempts to describe polar bonds in terms of Lewis structures.

Although, as proposed by Lewis, the octet rule is a purely empirical rule, the advent of orbital models appeared to add some theoretical support to the octet rule. For period 2 elements a maximum of only four orbitals, the 2s and the three 2p orbitals, are available for describing the bonds in terms of localized bonding and nonbonding orbitals, because other orbitals such as 3s, 3p, and 3d have energies that are too high. As a consequence, the octet rule came to be regarded more as a physical law than as a purely empirical rule. So it was assumed that hypervalent molecules are in some way special and have a type of bonding that is different from that in "normal" molecules. Because it acquired the status of a physical law, the octet rule exerted an important, but unjustified influence on ideas on bonding. For example, because the free noble gas atoms have an octet of electrons in their valence shell, it was for a long time believed that they were incapable of forming any compounds, although Pauling had suggested in the 1930s that they might form molecules such as XeF_6.

Not only molecules with LLPCN > 4, but all molecules of the elements in period 3 and beyond in their higher valence states, including most of their numerous oxides, oxoacids, and related molecules such as SO_3 and $(HO)_2SO_4$ should be regarded as hypervalent if AO bonds are described as double bonds (1). However, Lewis did not regard these molecules as exceptions to the octet rule because he wrote the Lewis structures of these molecules with single bonds and the appropriate formal charges (2).

(1)   (2)

For this reason the term "hypervalent" has often been restricted to the molecules of the elements of period 3 and beyond with LLCPN > 4. We have discussed the nature of AO bonds with A a period 2 element in Section 8.6, where we concluded that they are best represented as double bonds. We will later come to a similar conclusion with regard to AO bonds in which A is an atom of an element from period 3 and beyond. On this basis molecules such as $SO_2(OH)_2$ would be classified as hypervalent, as would the period 2 molecules OCF_3^- and ONF_3 as discussed in Chapter 8.

Originally hypervalent molecules were accommodated in the valence bond model by supposing that the 3d orbitals are available for bond formation by the period 3 elements in addition to the 3s and 3p orbitals. Five or six localized bonding orbitals could then be formed by the overlap of hybrid orbital such as sp^3d and sp^3d^2 with a suitable ligand orbital (Box 9.1). However, this description of the bonding was later questioned because of the relatively high energy of the 3d orbitals. Subsequently it was found in ab initio calculations that only very small contributions from d-type basis function are needed to obtain a minimum energy geometry in good agreement with experiment. As a consequence, it is often stated that d orbitals make only a small contribution to the bonding in hypervalent molecules.

It is difficult to give a localized orbital description of the bonding in a period 3 hypervalent molecule that is based only on the central atom 3s and 3p orbitals and the ligand orbitals, that is, a description that is consistent with the octet rule. One attempt to do this postulated a new type of bond called a three-center, four-electron (3c,4e) bond. We discuss this type of bond in Box 9.2, where we show that it is not a particularly useful concept. Pauling introduced another way to describe the bonding in these molecules, namely, in terms of resonance structures such as 3 and 4 in which there are only four covalent bonds. The implication of this description is that since there are only four cova-

▲ BOX 9.1 ▼

Hybrid Orbital Descriptions of the Bonding in Hypervalent Molecules

When the electron configurations of the elements were worked out, it became clear that the valence electrons of the period 2 elements must be accommodated in just four orbitals, the 2s and the three 2p orbitals. In the localized orbital model it is assumed that each bond can be described by a localized orbital formed by the overlap of one orbital on each of the bonded atoms. According to this model, therefore, a period 2 element can form bonds with at most four ligands so that electron configurations appeared to provide a justification for the octet rule.

Because the central atom in a hypervalent molecule of a period 3 element has more than four pairs of electrons in its Lewis structure, it was assumed that to describe the bonding in these molecules, it was necessary to use one or more of the 3d orbitals in addition to the 3s and 3p orbitals. The shapes of the five d orbitals are shown in Figure 1. The d orbitals can be combined with the s and p orbitals to give suitable spd hybrids with the correct relative orientation to correspond to the known geometry. Like sp hybrids, these orbitals are more localized in a particular direction than atomic orbitals and are therefore more suitable for forming localized bond orbitals. Each hybrid orbital is combined with a ligand atomic orbital to give a localized orbital corresponding to each of the bonds. The combination of one 3s, three 3p, and the $3d_{z^2}$ and $3d_{x^2-y^2}$ orbitals gives a set of six equivalent octahedral sp^3d^2 hybrid orbitals that are strongly directed toward the vertices of an octahedron and so are convenient for describing the bonding in molecules with an LLPCN of 6, which for main group elements always

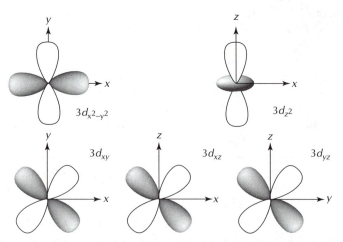

Figure 1. Conventional diagrams of the shapes of the 3d orbitals (Reproduced with permission from M. J. Winter, *Chemical Bonding*, 1994, Oxford Univ. Press, Oxford.

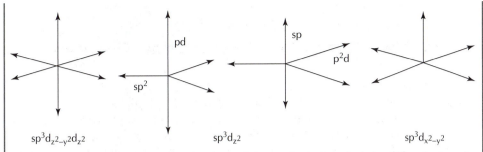

$sp^3d_{z^2-y^2}2d_{z^2}$ $sp^3d_{z^2}$ $sp^3d_{x^2-y^2}$

Figure 2. Diagrams showing the directions of the maxima of octahedral sp^3d^2 and trigonal bipyramidal and square pyramidal sp^3d hybrid orbitals.

have an octahedral geometry (Figure 2). However, other combinations of s, p, and two d orbitals correspond to other geometries. For example, the combination of the s, p, and $d_{x^2-y^2}$ and d_{xy} orbitals gives a set of spd hybrid orbitals with a trigonal prism geometry.

The combination of the s, three p, and the single d_{z^2} orbital gives a set of trigonal bipyramidal orbitals suitable for describing the bonding in molecules with LLPCN = 5, which with very few exceptions have a trigonal bipyramidal geometry. However, because the five vertices of a trigonal bipyramid are not equivalent, there is no unique set of equivalent sp^3d hybrid orbitals having this geometry. Rather, there is a complete range of possibilities ranging from two equivalent sd orbitals in the axial directions and three equatorial sp^2 orbitals in the equatorial directions to two equivalent axial sp orbitals and three equivalent equatorial spd_{z^2} orbitals (Figure 2). The first of these sets is usually chosen because its geometry corresponds better to the observed structure of these molecules in which the axial bonds are longer than the equatorial bonds, but this feature of the geometry is not predicted by the hybrid orbital model.

A set of five $sp^3d_{x^2-y^2}$ orbitals have a square pyramidal geometry (Figure 2). Because there is no way of deciding whether the bonding should be described by $sp^3d_{z^2}$ or $sp^3d_{x^2-y^2}$ hybrid orbitals without appealing to the experimentally determined geometry, we see again that the orbital model is only a method for *describing* the bonding in orbital terms, not a method for *predicting* the geometry.

The hybrid orbital model has been questioned because of the relatively high energy of the d orbitals, and it has been shown that, indeed, only a very small contribution from d-type basis functions is needed to obtain a good ab initio wave function. So it was concluded that it is not appropriate to utilize d atomic orbitals in an approximate atomic orbital based description of the bonding. As we discuss in Section 9.2 and in Box 9.2 alternative approximate descriptions of the bonding in hypervalent molecules can be given without making use of d orbitals. So the hybrid orbital model is no longer widely used, although it is still commonly found in more elementary textbooks.

▲ BOX 9.2 ▼
The Three-Center, Four-Electron (3c, 4e) Bond Model

The three-center, four-electron (3c, 4e) bond model, which is a combination of molecular orbital and localized orbital models, was proposed to avoid the use of d orbitals in the description of the bonding in hypervalent molecules. According to this model, each of the shorter equatorial bonds (153.4 pm) in PF_5 is described by localized orbitals formed from an sp^2 hybrid orbital on the phosphorus atom and a p orbital on a fluorine atom. The two longer axial bonds (157.7 pm) are together described by two of the three molecular orbitals formed from a single p orbital on phosphorus and an p orbital on each fluorine. The two lowest energy orbitals, namely, the bonding and nonbonding orbitals, are occupied each by a pair of electrons while the anti-bonding orbital remains empty (Figure 1).

In the nonbonding orbital two electrons are delocalized over the two fluorine atoms and do not contribute to the bonding, which is due only to the two electrons in the bonding orbital. This type of 3c, 4e bond is often denoted by a dashed line, as shown in Figure 2. Each P—F bond is effectively a half-bond, so this description of the bonding is roughly equivalent to the two resonance structures **1** and **2**:

$$F^- \quad P^+\!\!-\!\!-\!\!-F \qquad F\!\!-\!\!-\!\!-P^+ \quad F^-$$
$$(1) \qquad\qquad\qquad (2)$$

According to this description, each fluorine carries a formal charge of -0.5. This model of the bonding implies that the axial bonds are considerably weaker and longer than the equatorial bonds and that the charges on the axial fluorine ligands are much larger than those on the equatorial fluorine ligands. However, the difference in the bond lengths is smaller than this model implies, and as we shall see in Section 9.3 the charges

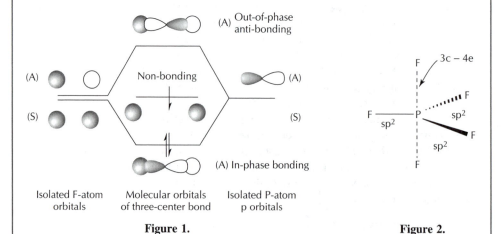

| Isolated F-atom orbitals | Molecular orbitals of three-center bond | Isolated P-atom p orbitals |

Figure 1.

Figure 2.

on the axial and equatorial fluorine atoms are very similar. The use of the term "three-center, four-electron" type of bond tends to suggest that it is new type of bond found only in hypervalent molecules, whereas it is only an alternative way of writing resonance structures which are themselves not an entirely satisfactory way of representing the bonding in hypervalent molecules.

The 3c, 4e bond has also been used in the description of the bonding in AX_6 molecules in which all the bonds are identical. In such cases it is necessary to resort to three resonance structures each of which has two 3c,4e bonds and two sp bonds (3, 4, and 5:

(3)   (4)   (5)

In this case the model becomes a rather complicated mixture of the 3c,4e model and the resonance model. There is therefore even less reason to use the model for describing the bonding in octahedral molecules such as SF_6 than there is for trigonal bipyramidal molecules such as PF_5. In summary the 3c,4e bond model gives only a very approximate and somewhat misleading description of the bonding in an AX_5 type of hypervalent molecule and is even less appropriate for describing other hypervalent molecules. The 3c,4e bond is not a special type of bond. It simply provides an unnecessarily complicated description of the bonding in hypervalent molecules in which there are five or more polar bonds, that is bonds formed by the unequal sharing of a pair of electrons. We note again the difficulties introduced by assuming that a bond line represents a purely covalent bond, that is an *equally* shared pair of electrons.

A satisfactory description of the bonding in hypervalent molecules can also be given in terms of molecular orbitals but this does not directly correspond to the very useful picture of five or more localized bonds (see, for example, Mingos, 1998, p. 250).

lent bonds, only one s and three p orbitals are needed in a valence bond or localized orbital description.

(3)   (4)

Structures such as these imply that the bonds are polar, which is consistent with ligands that are more electronegative than the central atom, as is often the case. When resonance structures are written in this way, it is assumed that the bond lines represent fully covalent (nonpolar) bonds, while the other bonds are fully ionic. A molecule such as SF_6 has six equivalent bonds, so we would need 15 structures such as **4** to fully describe the six equally polar bonds. This description implies a charge of +2.0 for the sulfur atom and a charge of −0.33 for each fluorine atom, in contrast to the calculated charges of +3.55 for sulfur and −0.59 for fluorine. So for a full and more accurate description of the bonding, we would have to add further resonance structures with larger charges and fewer covalent bonds. Clearly this type of description becomes very cumbersome and can be misleading. It is also clear that the choice of structures such as **4** just because they obey the octet rule is not justified.

Much of the confusion in describing the bonding in hypervalent molecules has arisen (1) because there is no generally accepted way of denoting the polar bonds in a structure and (2) because the octet rule has not always been fully understood. Lewis recognized that in most molecules the bonds are polar and the bonding electron pairs are not shared equally. Whether an electron pair is shared equally or unequally, Lewis denoted it by a bond line. If we do this, then resonance structures involving ionic bonds are not needed and **5** is an acceptable structure for SF_6:

(5)

In a Lewis structure a shared pair denoted by a bond line counts as contributing to the valence shell of both atoms, so that both atoms acquire an octet of electrons. Once we have introduced the concepts of a polar bond and unequal sharing of a pair of electrons, the meaning of the octet rule becomes less clear. The conventional Lewis structure of CF_4 (**6**) obeys the octet rule, but structures **7** and **8**, which would be used to describe the polarity of the bonds, do not.

(6)   (7)   (8)

The use of resonance structures such as **7** and **8** to describe bond polarity led to a subtle change in the meaning of the octet rule, namely, that an atom obeys the octet rule *if it does not have more than eight* electrons in its valence shell. As a result, resonance structures such as **7** and **8** are considered to be consistent with the octet rule. However, this is not the sense in which Lewis used the octet rule. According to Lewis, a structure such as **7** would not obey the octet rule because there are only three pairs of electrons in the valence shell of carbon, just as BF_3 does not obey the octet rule for the same reason. Clearly the octet rule as defined by Lewis is not valid for hypervalent molecules, which do, indeed, have more than four pairs of shared electrons in the valence shell of the central atom.

A qualitative molecular orbital description of a hypervalent molecule such as SF_6 which uses only the sulfur 3s and 3p orbitals in the construction of the molecular orbital, is described in several standard texts such as those by Mingos and by Huheey. According to this description, eight of the valence electrons occupy four bonding orbitals, while the remaining four occupy two nonbonding orbitals that are entirely composed of fluorine orbitals. This description is then equivalent to the resonance structures **4** and is similarly consistent with the polar character of the bonds. Although such a qualitative molecular orbital description of the bonding is useful, it is based on delocalized orbitals that cover the whole molecule. So it does not obviously accord with the well-defined and measurable properties such as length, energy, and force constant that can be attributed to the SF bonds. Moreover, as we shall see in Section 9.6, the AIM analysis of the electron density distribution shows that there are six bond paths or concentrations of charge density between the sulfur atom and the six fluorine atoms. SF_6 has six polar bonds and therefore does not obey the octet rule as defined by Lewis but the bonds are qualitatively the same as in CF_4 which does obey the octet rule.

We can summarize the foregoing discussion of the octet rule and hypervalent molecules as follows:

1. The octet rule is an empirical rule that is not as important as is often assumed, particularly in introductory textbooks.

2. Molecules of the elements of period 3 and beyond may have higher LLP coordination numbers than four, and therefore considered to be hypervalent, because their atoms are larger than those of the period 2 elements. In other words, more than a total of four ligands and lone pairs can pack around a central atom if it is from period 3 and beyond.

3. Hypervalency is not a consequence of some special type of the bonding. The bonds in hypervalent molecules are similar to those in any other molecules and may range from predominately ionic to predominately covalent.

4. The term hypervalent is not very useful and is misleading if it is taken to indicate some unusual or special type of bonding.

◆ 9.3 Bonding in the Fluorides, Chlorides, and Hydrides with LLP Coordination Number up to Four

In this section we discuss the bonding of the fluorides, chlorides, and hydrides of the elements of periods 3 and beyond with LLP coordination numbers up to four with particular emphasis on the elements of period 3. As might be expected these molecules show many similarities to the corresponding period 2 molecules, and the differences can be mainly attributed to the larger size and lower electronegativity of the atoms of a period 3 element compared to the corresponding period 2 element.

In the following discussion we will see again the usefulness of the calculated atomic charges in understanding the lengths and strengths and polar character of bonds. We will see also that the atomic charges determine the "size" of an atom as expressed both by its bonding radius and, as we have seen in discussing the LCP model (Chapter 4), by its ligand radius. Moreover, these atomic charges correlate well with qualitative expectations from electronegativities, and they give a quantitative expression to the qualitative concept of polar character.

9.3.1 Fluorides

Figure 9.1 shows contour plots of the electron density distributions for AlF_3, SiF_4, PF_3 and SF_2, and Figure 9.2 gives the corresponding plots of the Laplacian L. The strongly polar character of the bonds in AlF_3, SiF_4, and PF_3 is clearly evident in the almost spherical elec-

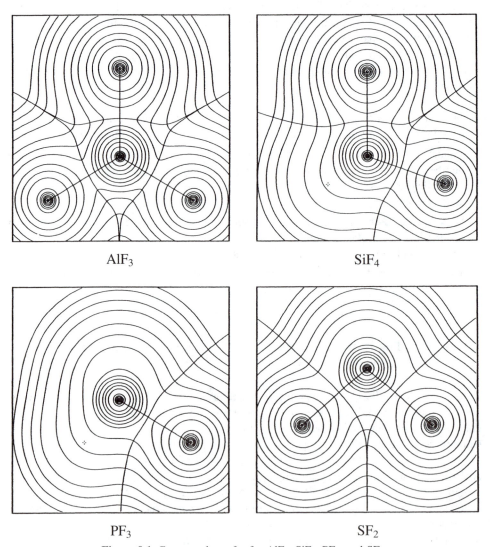

AlF₃   SiF₄

PF₃   SF₂

Figure 9.1 Contour plots of ρ for AlF_3, SiF_4, PF_3, and SF_2.

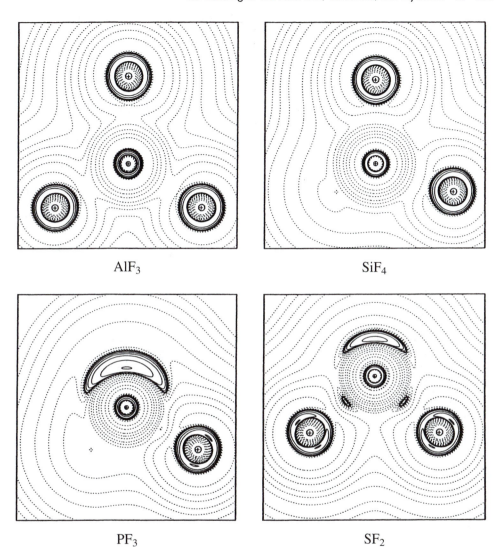

AlF$_3$

SiF$_4$

PF$_3$

SF$_2$

Figure 9.2 Contour plots of L for AlF$_3$, SiF$_4$, PF$_3$, and SF$_2$.

tron density around each nucleus. Because the bonding electron density in these molecules is largely transferred to the ligands, we do not see any bonding charge concentrations in the maps of L. The bulging of the electron density in the direction of the lone pairs can, however, be seen in the maps for PF$_3$ and SF$_2$, and the corresponding lone pair charge concentrations are quite evident in the maps of L. Only in SF$_2$ do we observe two small bonding CCs as well as the expected lone pair CCs. In the maps of L for PF$_3$ and SF$_2$, the rather weak localization of the nonbonding electrons of fluorine into a toroidal charge concentration can also be seen, as was the case for the fluorides of the period 2 elements.

Table 9.1 gives the calculated and experimental geometrical parameters, the bonding radii, and the atomic charges and bond critical point densities for the fluorides NaF, . . . , ClF and for some fluoro cations and anions of these elements. Figure 9.3 shows graphically how these properties vary across the period. The negative charge on fluorine decreases from NaF to ClF with

Table 9.1 Calculated and Experimental Geometrical Parameters and Calculated Atomic Charges and Bond Critical Point Densities for the Period 3 Fluorides

Molecule	Bond Lengths (pm)		Bond Angles (°)		ρ_b (au)	Atomic Charges		Radii (pm)	
	Calc.	Exp.	Calc.	Exp.		q(F)	q(A)	r_b(F)	r_b(A)
NaF	194.3	192.6	—	—	0.051	− 0.91	0.91	104.1	90.2
MgF$_2$	175.2	177	180	180	0.080	−0.88	1.76	96.6	79.1
AlF$_3$	163.9	163	120	120	0.115	−0.85	2.25	92.6	71.1
SiF$_4$	157.0	155.5	109.5	109.5	0.154	−0.81	3.26	91.2	65.8
PF$_3$	159.1	157	97.4	97.7	0.168	−0.76	2.28	93.9	64.9
SF$_2$	162.5	158.8	98.8	98.0	0.182	−0.58	1.16	94.2	68.3
ClF	166.5	162.8	—	—	0.187	−0.38	0.38	87.8	78.6
SiF$_3^+$	152.5	—	120	—	0.176	−0.73	3.21	88.2	64.3
AlF$_4^-$	169.9	165.8	109.5	109.5	0.096	−0.89	2.56	96.6	73.4
AlF$_6^{3-}$	189.7	181	90	90	0.056	−0.93	2.58	109.1	80.6
SiF$_6^{2-}$	172.9	169.4	90	90	0.101	−0.88	3.26	101.3	71.6

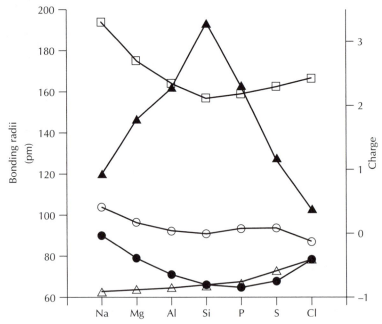

Figure 9.3 Atomic and bond properties of the period 3 fluorides: □, bond length; ○, r_bF; ●, r_bA; ▲, qA; △, qF.

the increasing electronegativity of the A atom, while the positive charge on A increases with the number of ligands up to SiF_4 and then decreases again as the number of ligands decreases. The bonding radii of A and F approximately follow the change in their charge, decreasing with decreasing negative charge and increasing positive charge. Because the change in the bonding radius of fluorine is rather small, the bond length decreases from NaF to SiF_4 and then increases again following the change in the bonding radius of A. As we pointed out in Chapter 2, the SiF bond, which has a length of 155.5 pm in SiF_4, is much shorter than the sum of the covalent radii (177 pm), which gives the length of a hypothetical nonpolar bond. This short length can be attributed mainly to the large positive charge on silicon, which considerably reduces the size of the silicon atom compared to the neutral atom. The bond critical point density ρ_b is less than ρ_b for the corresponding fluorides of the period 2 elements, consistent with the greater length of the bonds in the period 3 fluorides, and it increases continuously across the period, indicating that the covalent character of the bonding increases from NaF to ClF.

We see that the SiF bond, in particular, has both a strong covalent character as indicated by the high ρ_b value and a strong polar character as indicated by the large atomic charges. The large amount of density accumulated in the bonding region, as indicated by the large ρ_b value, and the large atomic charges are responsible for the great strength of the Si—F bond, which has an average bond enthalpy of 565 kJ mol^{-1}.

9.3.2 Chlorides

Contour maps of ρ and L for $AlCl_3$, $SiCl_4$, PCl_3 , and SCl_2 are given in Figures 9.4 and 9.5. Experimental and calculated bond lengths and bond angles for the chlorides of period 3 are given in Table 9.2. The variation in these properties across period 3 is shown in Figure 9.6. The atomic charges vary in the same way as for the corresponding fluorides, but they are smaller than in the corresponding fluorides (Table 9.1) because of the lower electronegativity of chlorine, while they are larger than those of the corresponding period 2 elements (Table 8.2) because of the smaller electronegativity of the period 3 elements. That these molecules are more covalent than the corresponding fluorides is also shown by the appearance of bonding charge concentrations in the contour maps of L for $SiCl_4$ and the following molecules. With increasing electronegativity of A, the bonding charge concentrations move from the ligand in $SiCl_4$ across the interatomic surface to the central atom in SCl_2.

The bonding radius of A decreases to silicon and then increases again following the change in its positive charge. The bonding radius of chlorine decreases continuously as its charge decreases. Consequently the A—Cl bond length decreases to Si and then is approximately constant, as the decrease in the bonding radius of Cl is roughly equal to the increase in the bonding radius of A. The bond critical point density increases steadily across the period as for the fluorides, but it is smaller than for the fluorides reflecting the greater length of the bonds.

9.3.3 Hydrides

Table 9.3 gives the calculated and experimental bond lengths and bond angles and the calculated atomic charges and bond critical point densities for the hydrides of the period 3 el-

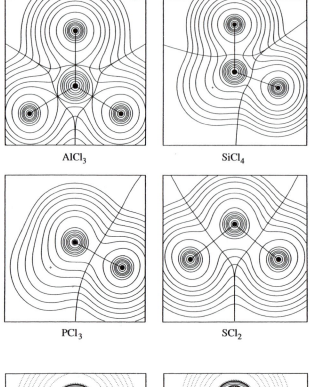

Figure 9.4 Contour plots of ρ for $AlCl_3$, $SiCl_4$, PCl_3, and SCl_2.

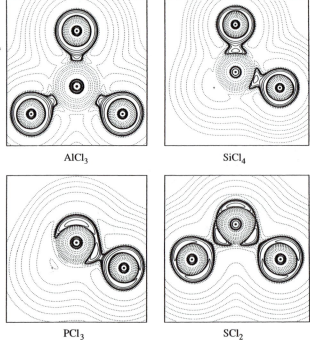

Figure 9.5 Contour plots of L for $AlCl_3$, $SiCl_4$, PCl_3, and SCl_2.

Table 9.2 Experimental and Calculated Bond Lengths and Bond Angles and Calculated Atomic Charges and Bond Critical Point Densities for the Period 3 Chlorides

Molecule	Bond Lengths (pm)		Bond Angles (°)		p_b (au)	Atomic Charges		Radii (pm)	
	Exp.	Calc.	Exp.	Calc.		q(Cl)	q(A)	r_b(Cl)	r_b(A)
NaCl	236	237.5	—	—	0.035	−0.87	+0.87	139.6	97.9
MgCl$_2$	—	217.8	180	180	0.057	−0.83	+1.65	131.7	86.2
AlCl$_3$	206.8	208.3	120	120	0.079	−0.78	+2.33	129.3	79.0
SiCl$_4$	202.0	204.3	109.5	109.5	0.106	−0.69	+2.77	128.9	75.4
PCl$_3$	204.0	209.0	100.4		0.120	−0.43	+1.29	120.1	88.9
SCl$_2$	201.5	206.1	102.7		0.133	−0.20	+0.40	108.9	97.2
Cl$_2$	199	203.4	—	—	0.140	0	0	101.7	101.7
AlCl$_4^-$		213.5	109.5	109.5	0.065	−0.84	+2.37	135.5	81.5
SiCl$_3^+$		196.5		120.0	0.085	−0.49	+2.44	119.2	77.4

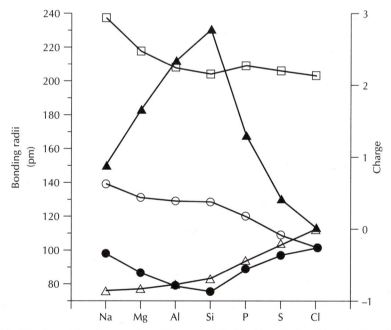

Figure 9.6 Atomic and bond properties of the period 3 chlorides: □, bond length; ○, r_bCl, ●, r_bA; ▲, qCl, △, qA.

ements. The variation of these properties across the period is shown in Figure 9.7. As we expect, the negative charge on hydrogen decreases rapidly across the period, becoming positive in HCl, while the charge on A increases from Na to Si and then decreases again. The bonding radii follow these trends closely. The change in the bonding radius of H is

Table 9.3 Calculated Bond Lengths and Bond Angles and Calculated Atomic Charges and Bond Critical Point Densities for the Hydrides of the Period 3 Elements

Molecule	Bond Lengths (pm)		Bond Angles (°)		p_b (au)	Atomic Charges		Radii (pm)	
	Exp.	Calc.	Exp.	Calc.		q(H)	q(A)	$r_b(A)$	$r_b(H)$
NaH	—	191.6	—	—	0.025	−0.82	+0.82	71.2	88.0
MgH₂	—	170.6	180	180	0.055	−0.81	+1.61	57.8	74.8
AlH₃	—	157.7	120	120	0.084	−0.79	+2.36	53.0	65.9
SiH₄	148.0	147.3	109.5	109.5	0.122	−0.72	+2.90	70.0	39.0
PH₃	142.1	140.5	93.5	95.4	0.165	−0.57	+1.69	73.5	28.1
SH₂	133.6	132.6	92.1	94.1	0.221	−0.14	+0.28	75.8	20.5
ClH	127.4	126.4	—	—	0.257	+0.22	−0.22	76.9	15.5

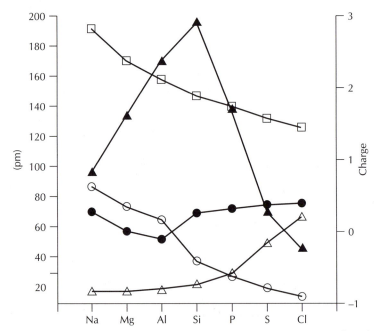

Figure 9.7 Atomic and bond properties of the period 3 hydrides: □, bond length; ○, r_bH, ●, r_bA; ▲, qA, △, qH.

so large as its charge decreases from −0.82 to −0.14 and then becomes positive (+0.22) that this is the major factor in causing the bond length to decrease continuously from NaH to ClH. The bond critical point density increases across the period as the bonding changes from predominantly ionic to polar as in SiH₄ and then to predominantly covalent in H₂S and HCl.

◆ 9.4 Geometry of the Fluorides, Chlorides, and Hydrides with LLP Coordination Number Up to Four

Figure 9.8 summarizes the geometrical data for some PX_3E, PX_3O, and PX_4^+ molecules and includes the corresponding NX_3E molecules for comparison. The bond angles are smaller in the PX_3E molecules than in the corresponding NX_3E molecules. These smaller angles are due to the larger size of the phosphorus atom and the consequent greater length of the PX bonds, which allows the ligands to subtend a smaller angle while remaining close-packed. The charge on fluorine in PF_3 is larger than in NF_3 so we expect the $X \cdots X$ distance to be larger than in NF_3, as is observed.

The data in Table 9.4 show that four-coordinated AX_4 molecules have longer bonds than the corresponding three-coordinated AX_3 molecules, as expected from the LCP model. However, the $X \cdots X$ distances are a little shorter in four- than in three-coordinated molecules and, as we shall see in Section 9.6, they are shorter still in six-coordinated molecules. Whereas the ligands in period 2 molecules can be treated as hard spheres, this appears not to be the case for molecules of the elements of period 3 and beyond. The bonds in period 2 molecules are shorter and stronger than in the corresponding period 3 molecules so that they are able to pull the ligands together until they reach the effective limit of their compressibility. Thus the interligand distance is essentially constant from molecule to molecule. However, the weaker longer bonds of the molecules of period 3 elements are unable to squeeze the ligands together as strongly and so they do not always reach equilibrium at the same $X \cdots X$ distance. In the limiting case of zero ligand–ligand repulsions, the bonds would all have the

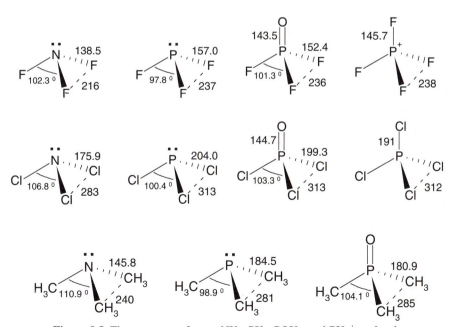

Figure 9.8 The geometry of some NX_3, PX_3, POX_3, and PX_4^+ molecules.

Table 9.4 Bond Lengths (pm) and Interligand Distances (pm) in AX_3 and AX_4 Molecules

Molecule	Bond Length	X···X	Molecule	Bond Length	X···X
AlF_3	163.0	282	AlF_4^-	165.8	270
$AlCl_3$	206.8	369	$AlCl_4^-$	217	354
AlH_3	157.7[a]	273	AlH_4^-	161	259
SiF_3^+	152.5[a]	264	SiF_4	155.5	253
$SiCl_3^+$	196.5[a]	340	$SiCl_4$	204.3	333
			SiH_4	141.4	242

[a]Calculated.

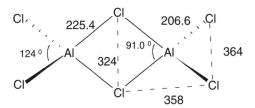

Figure 9.9 The structure of Al_2Cl_6.

same length independent of coordination number. In the other limiting case of the hard sphere model, the interligand distances would all be the same. For the molecules in Table 9.4 we see that the bond lengths increase a little from three- to four-coordination, while the ligand–ligand distances decrease a little, consistent with considerable repulsion between the ligands, which are however somewhat compressible.

Unlike BCl_3, $AlCl_3$ is very largely dimerized in the vapor state to Al_2Cl_6, which has the structure shown in Figure 9.9. Both aluminum atoms are approximately tetrahedrally coordinated. The terminal Cl bonds have lengths of 206.6 pm, but the bridging bonds are considerably longer (225.4 pm), which is a common feature of all bridging bonds. The attraction of the Cl atom by the two positively charged Al atoms pulls the Cl atoms together strongly enough that the Cl···Cl distance across the ring (324 pm) is shorter than the other Cl···Cl contacts (364 and 358 pm) and shorter than the contacts in $AlCl_4^-$ (348 pm). This short cross-ring Cl···Cl distance, and the correspondingly small ClAlCl angle of 91.0° allow the angle between the terminal Cl ligands to open up to open up to 124°. The same geometric feature is found in the analogous B_2H_6 molecule (Chapter 8).

9.4.1 Oxides, Hydroxides, and Other Oxomolecules

The oxides and hydroxides of Na, Mg, and Al are predominately ionic solids. The $Si(OH)_4$ molecule readily loses water to give SiO_2, which is a three-dimensional solid that has very strong bonds, presumably owing to large atomic charges and a high bond critical point density. Phosphorus, sulfur, and chlorine form a variety of hydroxides and oxides in most of their oxidation states, with those formed in their higher oxidation being generally the more common and more stable molecules. In general, the simple hydroxides of these elements such as $P(OH)_5$, $S(OH)_6$, and $Cl(OH)_7$ are unknown, whereas the oxohydroxo molecules that would be formed from them by loss of water such as $OP(OH)_3$, $O_2S(OH)_2$, and O_3ClOH are the well-known phosphoric, sulfuric, and perchloric acids in which the central atom is only four-coordinated. Presumably the formation of these four-coordinated molecules is favored by the reduction in ligand–ligand repulsions. Note that the larger iodine and tellurium atoms form the stable six-coordinated molecules $OI(OH)_5$ and $Te(OH)_6$.

Figure 9.8 gave geometrical data for some PX_4^+ and POX_3 molecules (X = F, Cl, CH_3) and includes also the data for the corresponding PX_3 and NX_3 molecules for comparison, all of which have a LLP coordination number of four. We see that the F$\cdots$F, Cl$\cdots$Cl, and C$\cdots$C distances are all essentially constant, consistent with the LCP model. From these distances we can deduce the ligand radii given in Table 9.5. The interligand distances in the phosphorus molecules are larger than those in the corresponding molecules of nitrogen because the charge on the halogen has a much larger negative value than in the corresponding nitrogen molecules where, because of the large electronegativity of nitrogen, fluorine has a small negative charge and chlorine and methyl have positive charges.

Replacing a Cl or an F ligand in PF_4^+ and PCl_4^+ by a doubly bonded oxygen considerably reduces the XPX bond angle and increases the length of the PX bonds, while the X$\cdots$X distance remains almost constant. In Chapter 8 we discussed the analogous effect of replacing a fluorine ligand in NF_4^+ and CF_4 with an oxygen ligand. According to the VSEPR model, an oxygen ligand has this effect because the PO bonding domain is larger than a PX domain. According to the LCP model, because the oxygen forms a shorter bond than either F or Cl, it repels these ligands more strongly than they repel each other, thus increasing the OPX angle and decreasing the XPX angles as explained for C=O bonds in Section 8.6. The charge on oxygen is much larger than that on the halogen ligands, which presumably is the main reason for the very short bond in all the POX_3 molecules. The four-coordinated molecules SO_2XY similarly have geometries that indicate that the ligands are close-packed. Ligand–ligand distances for some molecules of this type are given in Table 9.6. We see that these distances are remarkably constant for two given ligands, showing that the ligands are close packed. From these distances we can deduce the ligand radii for ligands attached to sulfur given in Table 9.5. These ligand radii are slightly smaller than the corresponding radii

Table 9.5 Ligand Radii (pm) for Period 3 Atoms with LLPCNs 4 and 6

	Central Atoms				
Ligand	Al	Si	P	S	Cl
F*	135	127	118	114	108
	(127)	(119)	(111)	(109)	
O		132	126	124	
C		154	142	136	
Cl	177	167	157	155	

*Values in parentheses are for LLPCN = 6. All the others are for LLPCN = 4.

Table 9.6 Interligand Distances (pm) in Some XYSO$_2$ Molecules

Molecule	O···F	O···Cl	O···C	O···O
F$_2$SO$_2$	239			245
Cl$_2$SO$_2$		280		250
FSO$_2$Cl	237	278		247
FSO$_2$OH	237			248
(HO)$_2$SO$_2$				250
MeOSO$_2$F	237			250
MeOSO$_2$Cl		278		249
MeSO$_2$F	238		260	248
MeSO$_2$Cl		281	261	248
Me$_2$SO$_2$			261	248
Mean	238	279	261	248

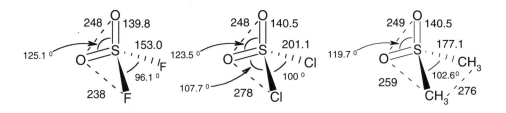

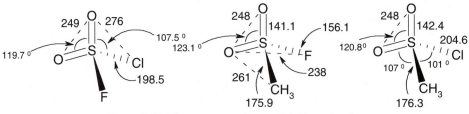

Figure 9.10 The geometry of some SO$_2$X$_2$ molecules.

for ligands attached to phosphorus, consistent with the expected slightly smaller negative charge on the ligands due to the greater electronegativity of phosphorus than sulfur. Bond lengths and bond angles for some of these molecules are given in Figure 9.10. We see that the OSO angle in these molecules is always larger than the other angles, as expected on the basis of both the VSEPR and LCP models.

♦ **9.5   Molecules with an LLP Coordination Number of Five**

Although they are much more common than was at one time believed, molecules of the non-metallic elements with an LLP coordination number of greater than four are nevertheless relatively rare. They are very largely limited to the halides, particularly the fluorides, chlorides, oxides, alkyls and aryls, in all of which the ligand atom (F, Cl, $=$O, and C) that is directly attached to the central atom is small. The other atoms in a polyatomic ligand such as CH_3 and CF_3 are sufficiently distant from neighboring groups that they have little effect on the bond angles. Similar molecules might be expected when the ligand atoms are O and N, as in hydroxides and amides. Such molecules are, however, not known, apparently because they are able to eliminate a stable small molecule such as H_2O or NH_3 to give a less crowded and therefore preferred four-coordinated molecule. Thus the fluorides PF_5 and SF_6 are known but $P(OH)_5$ and $S(OH)_6$ are not known because if formed they would presumably eliminate water to give the corresponding four-coordinate oxoacids H_3PO_4 and H_2SO_4. Similarly, the amide $P(NH_2)_5$ is not known, but there are many phosphorus(V)–nitrogen compounds in which phosphorus is four-coordinated. Common examples are the cyclic phosphazenes such as $(PNCl_2)_3$ (9).

(9)

Phosphorus pentaphenyl (PPh_5) is a stable trigonal bipyramidal molecule, but $P(CH_3)_5$ is not known, although the four-coordinate $Ph_3C=CH_2$ is a stable molecule. We can imagine that if $P(CH_3)_5$ were formed it would eliminate CH_4 to give $(CH_3)_3P=CH_2$, as presumably happens in the attempted preparation of $P(CH_3)_5$ by the reaction $(CH_3)_3PI$ with CH_3Li, which gives $(CH_3)_3P=CH_2$, CH_4, and LiI rather than $P(CH_3)_5$. However, the slightly larger As atom forms the stable pentamethyl $As(CH_3)_5$.

9.5.1 AX₅ Molecules

Except for $Sb(C_6H_5)_5$ and $InCl_5^{2-}$, all the known AX_5 molecules have trigonal bipyramidal geometries. Some examples are given in Table 9.7. $Sb(C_6H_5)_5$ and $InCl_5^{2-}$ have a square pyramidal geometry, which as we saw in Chapter 4 is close in energy to the trigonal bipyramidal geometry . The structures of $Sb(C_6H_5)_5$ and $InCl_5^{2-}$ were determined in the solid state, and their unusual geometry is probably due to intermolecular interactions. This supposition is supported by the observation that in crystalline $Sb(C_6H_5)_5 \cdot {}^1/_2 C_6H_6$ the antimony pentaphenyl molecule has a very nearly trigonal bipyramidal geometry. As we discussed in Chapter 5, five ligands or five bonding electron pairs cannot adopt a truly close-packed arrangement. If all the bonds were the same length, the equatorial ligands would not be close-packed. In fact, the axial bonds are longer than the equatorial bonds, although the ratio of the two lengths varies from one molecule to another (Table 9.7). The two axial ligands have three close neighbors at 90°, while the two axial ligands have two close neighbors at 90° and two more at 120°. If all the ligands were at the same distance from the nucleus, the total repulsion on the axial ligands would be greater than on the equatorial ligands, so this would not be an equilibrium geometry. Equilibrium can be reached only if the axial ligands move to a greater distance from the nucleus, allowing the equatorial ligands to move a little closer. So the axial bonds are always longer than the equatorial bonds, although $Bi(CH_3)_5$ appear to be an exception, since the axial

Table 9.7 Bond Lengths and Interligand Distances in AX₅ Molecules

Molecule	Bond Lengths (pm)		X···X (pm)	
	Axial	Equatorial	ax–eq	eq–eq
SiF_5^-	162.4	157.9	227	273
PF_5	157.7	153.4	220	266
PCl_5	212.7	202.3	294	350
AsF_5	171.1	165.6	238	287
$SbCl_5$	233.8	227.7	326	394
$Sb(CH_3)_5$	226	214	311	371
$Bi(CH_3)_5$	225	227	320	293

bonds are apparently very slightly shorter than the axial bonds. If the axial ligands are in contact with the equatorial ligands, then the equatorial ligands cannot be touching each other, so that, as we can see in Table 9.7, the equatorial–equatorial interligand distances are always greater than the axial–equatorial distances. The axial–equatorial interligand distances are close to those in the corresponding AX_6 molecules (Section 9.6), as we would expect with the axial ligands in contact with the equatorial ligands. Decreasing the number of fluorine ligands adjacent to a given fluorine from four in PF_6^- to three as in an axial bond in PF_5 reduces the total repulsion acting on this ligand, allowing the P—F bond to decrease in length from 159.5 pm in PF_6^- to 157.7 pm in PF_5, For the PF equatorial bonds, the number of close neighbors is reduced from four to two, and so their length decreases even further to 153.4 pm. The $F_{eq}···F_{eq}$ distance of 266 pm is much larger than the $F_{eq}···F_{ax}$ distance of 220 pm, showing that the equatorial ligands are not close-packed in this plane. They can be considered to be "touching" the axial ligands but are not "touching" each other.

The difference in the axial and equatorial bond lengths has been discussed in terms of the three-center, four-electron model (Box 9.2), but this model was postulated on the basis of the known difference in axial and equatorial bond lengths and so does not provide an explanation of this difference in bond lengths,

Figure 9.11 gives contour maps of ρ and L for PF_5, and the calculated bond lengths, atomic charges, and bond critical point electron densities are given in Figure 9.12. The calculated bond lengths agree well with the experimental values (Table 9.7) The nearly spherical contours around each of the atoms in the density map and the extremely small bonding charge concentrations in the maps of L show that these molecules are very polar; the large atomic charges also demonstrate this polarity. As might be expected, the charge on fluorine is close to that in PF_3 (Table 9.1), but the charge on phosphorus is considerably larger because of the larger number of F ligands.

The geometry of AX_5 molecules with more than one type of ligand is of considerable interest because there are two nonequivalent sites that the ligands may occupy. According to the VSEPR model, the most electronegative ligands, which have the smaller bonding domains, will occupy the more crowded axial sites. According to the LCP model, the smaller ligands will occupy the more crowded axial sites. Generally, more electronegative ligands are smaller than less electronegative ligands, so that both models usually lead to the same

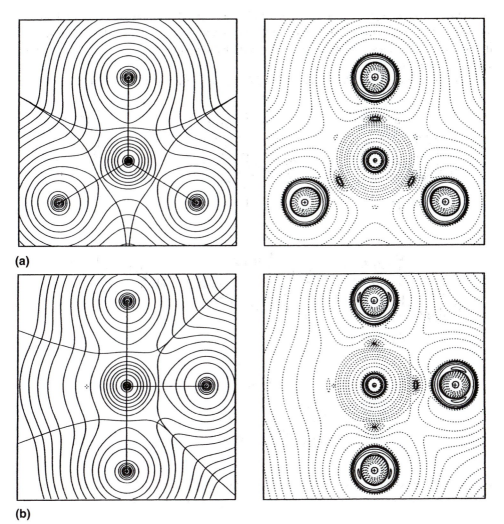

Figure 9.11 Contour maps of ρ and L for PF_5 (a) in the equatorial plane and (b) in the axial plane through an equatorial ligand.

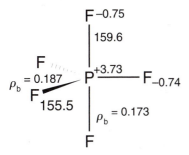

Figure 9.12 Geometry, atomic charges, and ρ_b values for PF_5.

predictions. For example, we see in Figure 9.13 that in the chlorofluorides of phosphorus the smaller and more electronegative fluorine ligands preferentially occupy the axial sites, and in the methyl fluorides the larger and less electronegative methyl ligands preferentially occupy the equatorial sites.

The molecules $(CF_3)_nPF_{5-n}$ and $(CF_3)_nPCl_{5-n}$ are of particular interest (Figure 9.14).

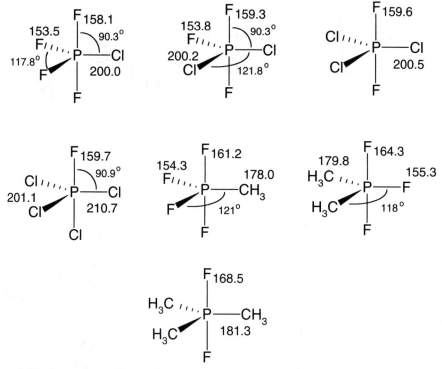

Figure 9.13 The geometry of some phosphorus chlorofluorides and some phosphorus methyl fluorides in which the larger and less electronegative ligands occupy the equatorial sites.

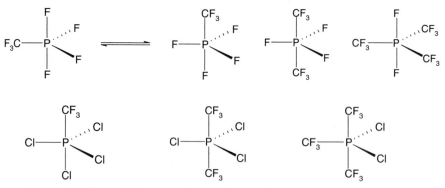

Figure 9.14 The stable isomers of some $PF_{5-x}(CF_3)_x$ and $(CF_3)_xPCl_{5-x}$ molecules.

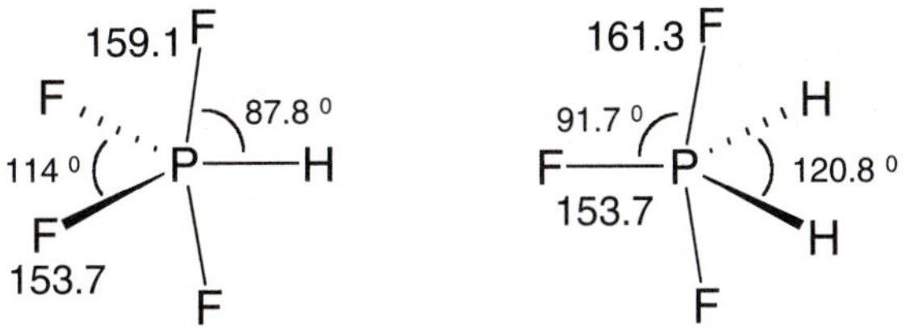

Figure 9.15 The geometry of the PF_4H and PF_3H_2 molecules.

A value of 3.5 has been given for the electronegativity of the CF_3 group, which is smaller than that of fluorine but larger than that of chlorine. So on this basis we expect the CF_3 ligand to preferentially occupy the axial sites in the $(CF_3)_nPCl_{5-n}$ molecules as observed, while we expect the CF_3 group to preferentially occupy the equatorial sites in the $(CF_3)_nPF_{5-n}$ molecules. However, the two possible isomers of CF_3PF_4 are in equilibrium, so they must have very similar energies. Although the carbon atom in a methyl ligand is larger than a fluorine ligand, the expected considerable positive charge on the carbon in the CF_3 group reduces its size so that it is probably comparable to that of a fluorine ligand. Consequently, there would be little difference in the preference of the two ligands for the axial and equatorial sites. If this is the case, it is not surprising that the two isomers have almost the same energy. The two CF_3 ligands $(CF_3)_2PF_3$ occupy the axial sites, whereas in $(CF_3)_3PF_2$ the two fluorine ligands are found in the axial sites. Although these geometries cannot be explained in detail, they are consistent with a lack of preference between the two ligands for the two sites.

A still more interesting case is presented by the two molecules F_4PH and F_3PH_2, in which hydrogen has a *smaller* electronegativity than fluorine but also a *smaller* size. In Figure 9.15 we see that the hydrogen ligand preferentially occupies an equatorial site in both molecules. This behavior is consistent with the lower electronegativity of hydrogen than fluorine but not with its smaller size. Still more curiously, the axial F ligands are bent *toward* the equatorial H ligands in both molecules, but the equatorial fluorine ligands are bent *away from* the equatorial H ligand in PF_4H while in F_3PH_2 the equatorial bond angles hardly deviate from 120°. These apparently contradictory observations could be explained in terms of the VSEPR model if the P—H bonding domain were strongly flattened in the equatorial plane, but more studies will be needed to confirm this suggestion or to provide an alternative explanation.

9.5.2 AX₄═Y Molecules

In AX_4═Y molecules in which Y is a doubly bonded ligand, Y always occupies an equatorial site, consistent with the greater size of a double-bond domain and the generally shorter length of A═O bonds. Some examples of AX_4═Y molecules are shown in Figure 9.16a. The distortions of the bond angles from the ideal trigonal bipyramid angles are also consistent with the greater size of a double bond domain than a single-bond domain and the shorter A═O bond length. Both the axial bonds and the equatorial bonds are bent away from the ═O ligand. As the electronegativity of the ligand decreases from OSF_4 to H_2CSF_4 and the bonding electrons move closer to the sulfur atom, the angles between the F ligands decrease.

One aspect of the geometry of these molecules that is of particular interest is that the CH_2 and NH groups are in the axial plane. This geometry is consistent with a bent-bond

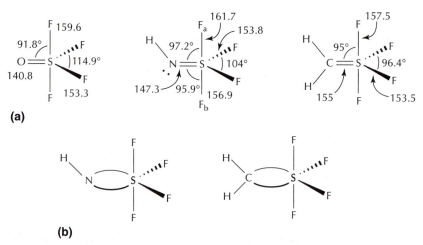

(a)

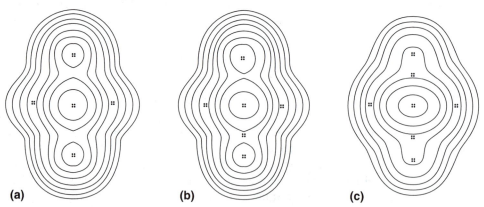

(b)

Figure 9.16 (a) The geometry of OSF$_4$, HNSF$_4$, and H$_2$CSF$_4$. (b) Bent-bond models of OSF$_4$ and HNSF$_4$.

model of the double bond in which the two components of the double bond lie in the equatorial plane, forming an approximately octahedral arrangement of six bonds around the sulfur atom (Figure 9.16b). This model is particularly appropriate for H$_2$C=SF$_4$, for which the bond lengths and bond angles approach those of a regular octahedral molecule. Figure 9.17 shows the electron density distribution in a vertical plane through the bond critical point of the Y=S bond for each of these three molecules: the density has an almost cylindrically symmetric shape for the S=O bond, as would be expected, but becomes increasingly elliptical in HN=SF$_4$ and H$_2$C=SF$_4$. Finally, we note that there is an appreciable difference in the length of the two axial SF bonds in HN=SF$_4$. This bond length

Figure 9.17 Contour maps of the electron density ρ in a plane through the bond critical point of the double bond in the molecules (a) O=SF$_4$, (b) HN=SF$_4$, and (c) H$_2$CSF$_4$.

difference is consistent with our earlier conclusion (Box 8.3) that the electron density is greater in the direction between two bonds than in the opposite lone pair direction, so that the nitrogen atom has a greater ligand radius in the direction of F_a than in the direction of F_b. Accordingly the SF_a bond is longer than the SF_b bond and NSF_a angle is larger than the NSF_b angle.

9.5.3 AX₄E, AX₃E₂, and AX₃E₂ Molecules

In AX₄E, AX₃E₂, and AX₃E₂ molecules the lone pairs, like multiply bonded ligands, invariably occupy the equatorial positions, giving the AX₄E molecules a disphenoidal geometry, the AX₃E₂, molecules a T-shaped geometry, and the AX₂E₃ molecules a linear geometry, consistent with the greater size of a nonbonding pair domain than a bonding pair domain. Some examples are given in Figures 9.18 and 9.19. In all these molecules the axial bonds are longer than the equatorial bonds. The larger size of the lone pair domains causes the observed distortion of the bond angles from the ideal values of 90° and 120°.

It is interesting that among molecules with an LLPCN of five or larger, only those, such as XeF₂, with an LLPCN of five are found to have as many as three lone pairs. This is pre-

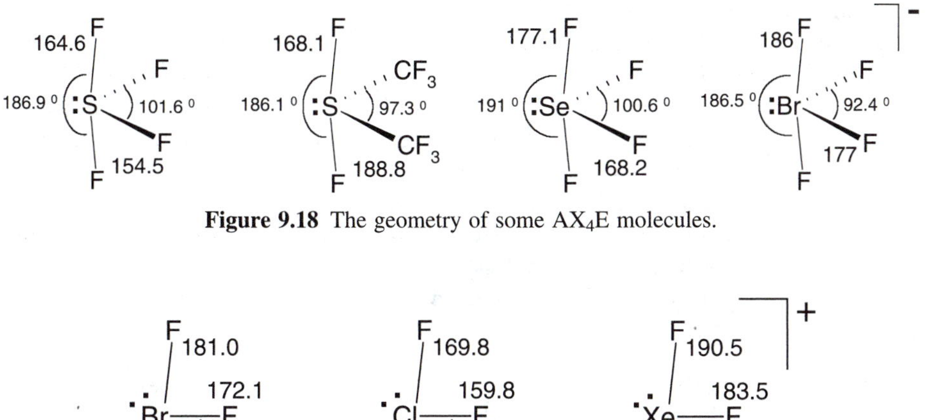

Figure 9.18 The geometry of some AX₄E molecules.

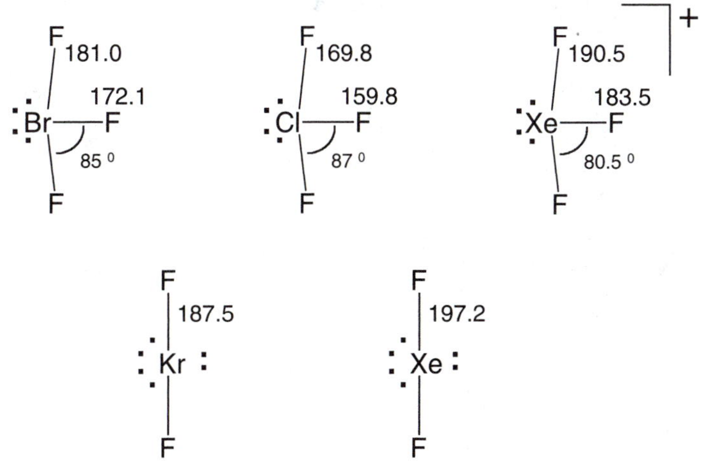

Figure 9.19 The geometry of some AX₃E₂ and AX₂E₃ molecules.

sumably because only in the equatorial positions of a trigonal bipyramid is sufficient space available to accommodate three lone pair domains. In molecules with an LLP coordination number of six or higher the six or more domains make angles of 90° or less with each other so that there is not sufficient room for three large lone pair domains.

◆ 9.6   Molecules with LLP Coordination Number Six

All main group molecules having an LLP coordination number of six have structures based on the octahedral arrangement of six domains and include AX_6, AX_5E, and AX_4E_2 molecules. They are restricted to molecules with small electronegative ligands, in particular, F, OH, N (in NF_2) and C (in CH_3 and CF_3), and Cl. Although NF_2 and CF_3 might be regarded as rather large groups, the F atoms in adjacent groups are considerably further apart than the carbon atoms, which therefore determine the effective size of the ligand and permit it to be present in a six-coordinated molecule. The smallest of these ligands, namely fluorine, is by far the most common ligand in AX_6 molecules. Examples of AX_6 molecules are provided by (1) anions of group 15 elements such as PF_6^-, PCl_6^-, SbF_6^-, and $Sb(OH)_6^-$, (2) neutral molecules of group 16 elements such as SF_6, SeF_6, TeF_6, $Te(OH)_6$, and various substituted SF_6 molecules such as SF_5Cl, SF_5CF_3, and SF_5NF_2, and (3) cations of group 17 elements such as and ClF_6^+, BrF_6^+, and IF_6^+.

In Section 9.2 we discussed several descriptions of the bonding in SF_6. Figure 9.20 gives contour plots of ρ and L for the SF_6 molecule. There is a bond path between the sulfur atom and each of the fluorine ligands, so there are six S—F bonds. These bonds have a considerable ionic character shown by the large atomic charges and the small bonding charge concentrations seen in the contour map of L, indicating that much of the electron density in the valence shell of the sulfur atom has been transferred to the fluorine ligands.

AX_5E molecules with one lone pair and $AX_5{=}Y$ molecules have a square pyramidal geometry in which the XAX angles are less than 90°. Some examples are given in Figure 9.21. There is an interesting, but unexplained, difference between the effect of a lone pair

Figure 9.20 Contour plots of ρ and L for the SF_6 molecule.

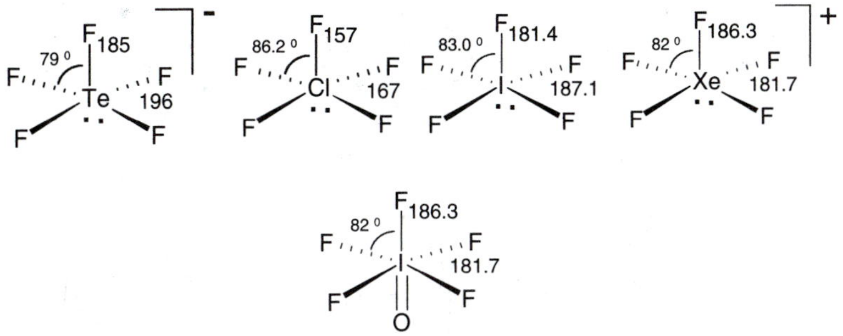

Figure 9.21 The geometry of some AX_5E molecules and of IOF_5.

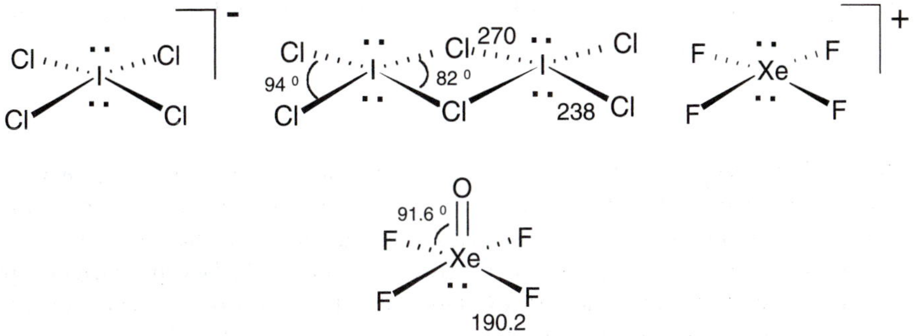

Figure 9.22 The geometry of some AX_4E_2 molecules and of $XeOF_4$.

and a double bonded ligand on the AX bond lengths. Although both decrease the XAX bond angles, a lone pair increases the length of the adjacent AX bonds relative to the trans bond while a doubly bonded oxygen increases the length of the *trans* AX bond relative to the adjacent bonds, at least in the single example of IOF_5.

When there are two lone pairs, or a lone pair and a doubly bonded ligand, they always occupy *trans* positions so as to minimize the interaction between them, as we see in Figure 9.22. So AX_4E_2 molecules have a square geometry. Examples include the ions ICl_4^-, ClF_4^-, BrF_4^-, IF_4^-, and the dimeric molecule I_2Cl_6. In this molecule the bridging bonds are longer than the terminal bonds, as we have seen for Al_2Cl_6 and as is found for all doubly bridged molecules. $AX_4{=}YE$ molecules have a square pyramidal geometry, and the single known example is the molecule $XeOF_4$, in which the fluorine ligands are bent slightly away from the oxygen ligand.

◆ **9.7 Molecules with an LLP Coordination Number of Seven or Higher**

9.7.1 AX₇ Molecules

Calculations based on the $1/r^n$ ($n = 6$) repulsion law for the points-on-a-sphere model in which the points can represent either electron pairs or ligands have shown that the pentago-

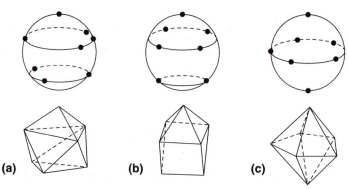

Figure 9.23 The three lowest energy arrangements for seven points on the surface of a sphere assuming a $1/r^6$ repulsion law: (a) the capped octahedron, (b) the capped trigonal prism, and (c) the pentagonal bipyramid.

nal bipyramid, the capped octahedron, and capped trigonal prism (Figure 9.23) have almost identical energies and can be interconverted in a process with essentially zero activation energy (Kepert, 1986). For lower values of n the pentagonal bipyramid has a slightly lower energy, while for $n = \infty$ (the hard sphere model), as well as for $n = 6$, the capped trigonal prism has the lowest energy. However, as for AX_5 molecules, the points-on-a-sphere model is not valid because seven points, like five points, cannot all be equivalent, and if they are not constrained to lie on the surface of a sphere, some move closer and some move further from the center of the sphere to attain a lower energy arrangement. So there appears to be no simple way to predict the lowest energy geometry. In fact, all the known main group examples of this type of molecule are based on the pentagonal bipyramidal geometry.

Molecules of this type are found only for the large atoms of the elements of periods 5 and beyond and in particular for Te, I, and Xe. Only the smallest ligands, namely, —F, —OMe, and =O, have been found in these molecules, suggesting that seven coordination is not possible with larger ligands. The structures of TeF_7^-, $MeOTeF_6^-$, $(MeO)_2TeF_5^-$, IF_7, and IF_6O^- have been determined experimentally by X-ray crystallography or electron diffraction. They are shown in Figure 9.24. They all have the pentagonal bipyramidal geometry with shorter axial than equatorial bonds, consistent with axial positions being less crowded than the equatorial positions. The axial positions have all their neighbors at ~90°, whereas each equatorial ligand has two neighbors at ~72°. The doubly bonded O ligand and the OMe group are found in a less crowded axial site because the oxygen ligand radius is larger than the fluorine ligand radius. The molecule $TeOF_6^{2-}$ is also known and has been shown by NMR spectroscopy to have a pentagonal bipyramidal geometry, but its structure has not been fully determined. The equatorial bonds are not exactly the same length in all the molecules, and they lie slightly above and below the equatorial plane, giving a slightly puckered ring of five fluorine atoms, suggesting strongly that the ligands in these molecules are indeed close-packed. This distortion of the equatorial ligands from a pentagonal planar geometry is not shown in Figure 9.24.

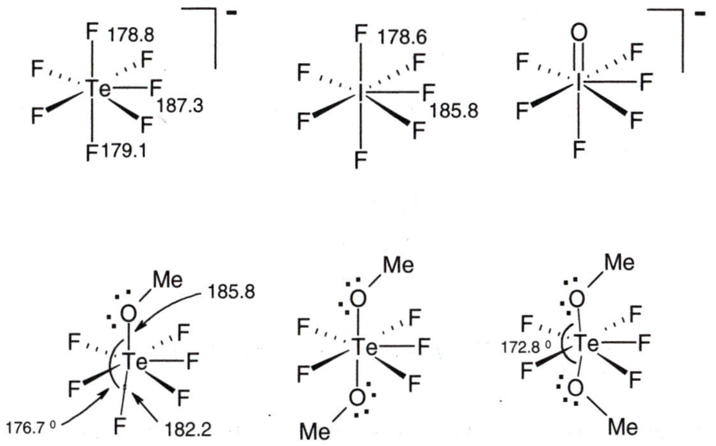

Figure 9.24 The pentagonal bipyramidal geometry of some AX_7 molecules.

When there are seven identical or very similar ligands as in IF_7, TeF_7^-, $Te(OMe)F_6^-$ and $Te(OMe)_2F_5^-$, NMR spectroscopy shows that the molecules are fluxional in solution: the equatorial and axial ligands change places, probably through a capped octahedral transition state with C_{3v} symmetry. The IOF_6^- molecule is not fluxional because the transition state is expected to be of considerably higher energy, since the $=O$ ligand would be in a more crowded equatorial site.

In the crystal structure of the $(CH_3)_4N^+$ salt of $CH_3OTeF_5^-$ the $OTeF_{ax}$ angle is 176.7°. In the crystal structure of the $(CH_3)_4N^+$ salt of $Te(OMe)_2F_5^-$ the anion is found in both the *syn* and *anti* forms (Figure 9.24) with an OTeO angle of 180.6° in the *anti* form and of 172.8° in the syn form. The significant deviation of the angle between the axial ligands from 180° in $Te(OMe)F_6^-$ and in syn-$Te(OMe)_2F^-$ is consistent with the unsymmetrical electron density distribution around the oxygen atom in the OMe group. As we discussed in Box 8.3, the density is greater on the bonded side of the oxygen atom than on the lone pair side so that the repulsion exerted by the equatorial ligands is greater on the bonded side than on the nonbonded side. Consequently, the ligand–ligand distance between the oxygen and the equatorial ligands is slightly smaller on the lone pair side of the oxygen atom than on the bonded side, producing the small difference in the angles between the axial ligands.

9.7.2 AX₆E Molecules and Sterically Weak and Inactive Lone Pairs

The known molecules of type are SeF_6^{2-}, BrF_6^-, IF_6^-, and XeF_6 (Figure 9.25), together with the ions SnX_6^{4-}, PbX_6^{4-}, SbX_6^{3-}, BiX_6^{3-}, SeX_6^{2-}, and TeX_6^{2-}, where $X = Cl$, Br, or I. Of these molecules, SeF_6^{2-}, IF_6^{2-}, and XeF_6 have distorted octahedral geometry with C_{3v} symmetry, while the others are apparently octahedral. If the nonbonding pair of electrons were fully sterically active, the AX_6E molecules would be expected to have a geometry based on the pentagonal bipyramid with a lone pair in a less crowded axial

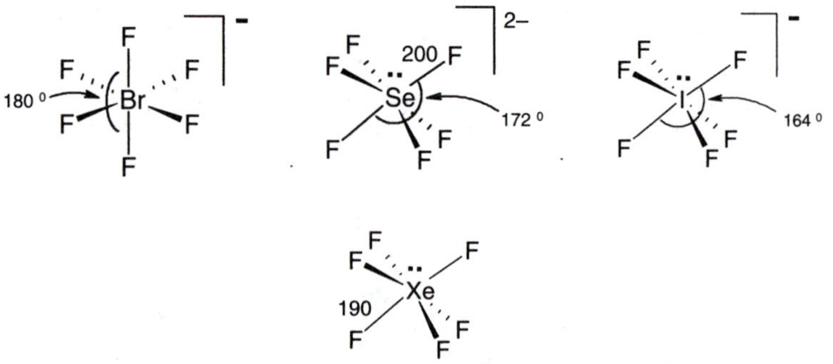

Figure 9.25 The geometry of some AX$_6$E molecules.

position. None of the molecules have this geometry. In the molecules with octahedral geometry, the lone pair apparently has no effect on the geometry and is said to be sterically inactive.

The molecules SeF$_6{}^{2-}$, IF$_6{}^-$, and XeF$_6$ have a C_{3v} geometry in which the lone pair appears to enlarge one of the faces of the octahedron and is said to be weakly stereochemically active. The degree of distortion from octahedral geometry increases from BrF$_6{}^-$ to SeF$_6{}^{2-}$ to IF$_6{}^-$, as shown by the angle between opposing ligands, which decreases from 180° in BrF$_6{}^-$ to 174° in SeF$_6{}^{2-}$ to 164° in IF$_6{}^-$. The bonds that surround the apparent position of the lone pair are longer and the angles between them are larger than on the opposite side of the molecule. Although the bonds vary somewhat in length as a consequence of interionic interactions in the solid state, the average length of the bonds surrounding the apparent position of the lone pair in SeCl$_6{}^{2-}$ is 200 pm, while the average length of the opposite bonds is 186 pm. In the ^{19}F NMR spectrum of XeF$_6$ all six fluorine ligands appear to be equivalent showing that XeF$_6$ is a highly fluxional molecule. We can imagine a process in which the unshared pair moves from one face of the octahedron to another, distorting each face in turn.

It seems that if the ligands are sufficiently large, as when they are Cl, Br, and I, or the central atom is small enough, as in BrF$_6{}^-$, six of them form a close-packed octahedral arrangement around the central atom so that there is no space available for a lone pair. In these molecules the two nonbonding electrons remain around the core and do not form a localized lone pair in the valence shell. We can think of these molecules as predominately ionic so that in SeCl$_6{}^{2-}$, for example, there is a central Se^{4+} ion surrounded by six close-packed Cl$^-$ ions. The Cl$\cdots$Cl distance in this molecule is 340 pm, giving a Cl radius of 170 pm, which is smaller than the ionic radius of 181 pm and consistent with the expectation that the charge on Cl will be slightly less than -1. Moreover, in all these molecules the bonds are unexpectedly long. For example, the SeCl bond in SeCl$_6{}^{2-}$ has a length of the 240 pm compared to 215.7 pm in SeCl$_2$. This long bond is consistent with a model in which Cl ligands with a charge of close to -1 are packed around a central Se core with a charge close to $+4$ rather than close to $+6$ as would be appropriate if the nonbonding pair was sterically active (Figure 9.26). We can regard BrF$_6{}^-$ in the same way. The bonds in BrF$_6{}^-$, which have an average length of 183 pm, are longer than

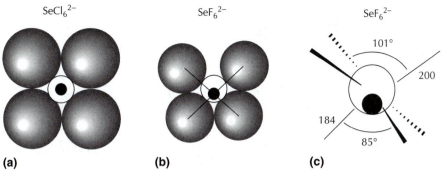

(a)   **(b)**   **(c)**

Figure 9.26 (a) Close packing of four of the Cl ligands in the $SeCl_6^{2-}$ ion around a central Se_4^{2+} "core" consisting of the Se^{6+} core and two nonbonding electrons. (b) Close packing of four of the F ligands in the SeF_6^{2-} around the Se^{4+} core, showing the distortion produced by the slight protrusion of the domain of the two nonbonding electrons of the Se^{4+} core into the valence shell. (c) The C_{3v} distorted octahedral geometry of the SeF_6^{2-} ion. In (a) and (b) the large spheres are Cl^- and F^-, respectively, the small black dots are Se^{6+}, and the larger open circles represent the spherical or slightly distorted domain of the 2 nonbonding electrons.

both the axial and equatorial bonds in BrF_5, which have lengths of 170 and 177 pm, respectively.

The molecules SeF_6^{2-}, IF_6^-, and XeF_6 can be regarded as intermediate between the octahedral molecules above and a molecule in which there are six ligands and a sterically active lone pair. Because there is only a small amount of additional space available in the valence shell, the nonbonding pair of electrons can be imagined as being extruded into the valence shell only to a small extent, remaining largely delocalized around the central core. Hence the nonbonding electrons produce only a small distortion of the geometry from octahedral, increasing the angle between the adjacent fluorine ligands and increasing the length of these bonds, as we saw in Figure 9.20.

9.7.3 AX₅E₂ Molecules

The molecule XeF_5^{2-} is the only known example of the AX_5E_2 type of molecule. It has the expected planar pentagonal geometry based on the pentagonal bipyramid with a lone pair in each of the less-crowded axial positions and an Xe—F bond length of 201 pm (Figure 9.27).

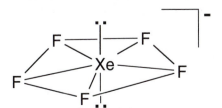

Figure 9.27 The AX_5E_2 pentagonal bipyramidal geometry of the XeF_5^- ion.

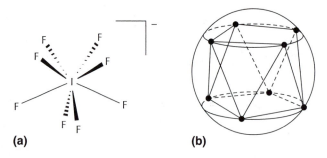

Figure 9.28 (a) Points-on-a-sphere model of a square antiprism. (b) The square antiprism geometry of the IF_8^{2-} ion.

(a)   **(b)**

9.7.4 AX_8 and AX_8E Molecules

IF_8^- is the only known example of an AX_8 molecule among the main group elements and XeF_8^{2-} the only known example of an AX_8E molecule. According to the points-on-a-sphere model, the square antiprism is the lowest energy geometry for any value of n in the $1/r^n$ force law (Figure 9.28). The structures of the IF_8^- and XeF_8^{2-} ions have been determined by X-ray crystallography of several of their salts. They both have an almost perfect square antiprism geometry with very nearly equal bond lengths and bond angles. IF_8^- is an example of an AX_8 molecule, while XeF_8^{2-}, which has an unshared pair of electrons and is an example of an AX_8E molecule, provides another example of a sterically inactive lone pair. The steric inactivity of the unshared pair of electrons is not surprising in view of the high coordination number. The Xe—F bonds in XeF_8^{2-} have an average length of 202 pm, which is considerably longer than the I—F bonds in IF_8^{2-} (average length of 189 pm). This difference in bond length is consistent with the supposition that the nonbonding pair of electrons has a spherical distribution surrounding an Xe^{8+} central core and does not form a localized lone pair in the valence shell.

The steric activity of a lone pair decreases as the number of ligands increases and the steric crowding of the ligands increases, correspondingly. For example, the nonbonding pair of electrons exerts a full steric effect in XeF_5^+ which has a square pyramidal geometry, a weak effect in XeF_6, which has a C_{3v} distorted octahedral geometry, and no effect in XeF_8^{2-}. The same trend is observed when the number of ligands remains the same but the size of the central atom is reduced. For example, IF_6^- and XeF_6 have C_{3v} distorted octahedral structures with the lone pair occupying the capping position in a capped octahedron, but BrF_6^- has a regular octahedral structure with a sterically inactive nonbonding pair. Increased steric crowding resulting from an increased size of the ligand also reduces the steric effect of a nonbonding electron pair. For example, whereas SeF_6^{2-} has a C_{3v} distorted octahedral geometry, $SeCl_6^{2-}$ has a regular octahedral geometry.

◆ 9.8 Molecules of the Transition Metals

In contrast to the nonmetals of the main group, elements the transition metals form only a relatively few compounds that are composed of simple isolated molecules, although they form many complex ions that exist as crystalline solids with an appropriate counter anion.

The tetrahedral and octahedral geometries are particularly common for these molecules and complex ions, whereas the shapes produced by the presence of lone pairs in the valence shell of a main group nonmetal such as the trigonal pyramidal AX_3E shape, and the disphenoidal AX_4E shape, are not common. These observations suggest that close packing of the ligands is a very important factor in determining the geometry of many molecules and complex ions of the transition metals. But perhaps the most important reason for the common occurence of the AX_n shapes, and other shapes not observed for nonmetal molecules, and the comparative rarity of the AX_nE_m shapes is that unpaired electrons are not in general found in the valence shell of the transition metal atoms in their molecules. Rather they are found in the outer shell of the core. For example, in the VCl_3 molecule the two unpaired nonbonding electrons occupy the incompletely filled $n = 3$ shell which lies mainly inside the valence shell of the vanadium atom. These two unpaired electrons cause only a small distortion of the spherical shape of the core and a correspondingly very small distortion of the planar triangular D_{3h} geometry expected on the basis of ligand-ligand repulsions. The calculated geometry of the ground state is planar with C_{2v} symmetry rather than D_{3h} or C_{3v} as for an AX_3E molecule. the molecule has one bond angle of 129° and two of 115° and one bond of length 176.8 pm and two bonds of length 174.8 pm. (Solomonik et al.). This distortion has, however, not been observed experimentally as it is obscured by the vibrational motion of the molecule at ordinary temperatures. The molecule MnF_3 with four unpaired electrons in the core is distorted in the same way but more substantially so that it has a very nearly T-shape geometry. An electron deffraction study showed that it has one bond of length 172.8 pm and two of length 175.4 pm and one bond angle of 143.3° and two of 106.4° with which the calculated structure is in good agreement (Hargittai et al.).

Although the ligand field theory can be used to rationalize the geometry of some transition metal molecules and complex ions, the study of the shapes of transition metal molecules in terms of the electron density distribution is still the subject of research and it has not reached a sufficient stage of development to enable us to discuss it in this book.

▶ Further Reading

F. A. Cotton, G. Wilkinson, C. A. Murillo, and M. Bochman, *Advanced Inorganic Chemistry,* 6th ed., 1999, Wiley, New York.
    A comprehensive reference book that gives information on structure and also on the preparation and properties of inorganic substances. Earlier editions are still a useful source of information particularly on bonding models.
R. J. Gillespie and I. Hargittai, *The VSEPR Model of Molecular Geometry,* 1991, Allyn & Bacon, Boston.
    Chapter 5 describes the VSEPR model for the main group period 3 elements.
N. N. Greenwood and A. Earnshaw, *Chemistry of the Elements,* 1984, Pergamon Press, Oxford.
    A comprehensive reference book, particularly strong on the main group elements.
I. Hargittai, *The Structure of Volatile Sulphur Compounds,* 1985, Reidel, Dordrecht.
    A comprehensive discussion of the geometry of molecules with a central sulfur atom in which considerable use is made of the VSEPR model.
J. E. Huheey, E. A. Keiter, and R. L. Keiter *Inorganic Chemistry,* 4th ed., 1993, HarperCollins, New York.

D. Kepert, *Inorganic Stereochemistry,* 1986, Springer-Verlag, New York.
   A detailed discussion of geometry in terms of the points-on-a-sphere model.
D. M. P. Mingos, *Essential Trends in Inorganic Chemistry,* 1998, Oxford University Press, Oxford.
N. C. Norman, *Periodicity and the p-Block Elements,* 1994, Oxford University Press, Oxford.
D. F. Shriver, P. W. Atkins, and C. H. Langford, *Inorganic Chemistry,* 1990, Freeman, New York.

▶ References

These papers discuss molecules with nonstereochemically active and weakly active lone pairs.

L. S. Bartell, R. M. Gavin, H. B. Thompson, and C. L. Chernick, *J. Chem. Phys. 43,* 2547, 1965.
R. J. Gillespie and E. A. Robinson, *Adv. Mol. Struct. Res. 4,* 1, 1998.
M. Kaupp, C. van Willen, R. Franke, F. Schmitz, and W. Kutzelnigg, *J. Am. Chem. Soc. 118,* 11939, 1996.
A.-R. Mahjoub, X. Zhang, and K. Seppelt, *Chem. Eur. J. 1,* 281, 1995.
E. A. Robinson, G. L. Heard, and R. J. Gillespie, *J. Mol. Struct. 485–486,* 305, 1999.

These papers describe molecules with LLPCN > 6.

S. Adam, A. Ellern, and K. Seppelt, *Chem. Eur. J. 2,* 398, 1996.
K. O. Christe, E. C. Curtis, D. A. Dixon, H. P. Mercier, J. C. P. Saunders, and G. J. Schrobilgen, *J. Am. Chem. Soc. 113,* 3351, 1991.
K. O. Christe, D. A. Dixon, A. R. Mahjoub, H. P. A. Mercier, J. C. P. Saunders, K. Seppelt, G. J. Schrobilgen, and W. W. Wilson, *J. Am. Chem. Soc. 115,* 2696, 1993.
K. O. Christe, D. A. Dixon, J. C. P. Saunders, G. J. Schrobilgen, S. S. Tsai, and W. W. Wilson, *Inorg. Chem. 34,* 1868, 1995.
A.-R. Mahjoud, T. Drews, and K. Seppelt, *Angew. Chem. Int. Ed. Engl. 31,* 1036, 1992.

These papers describe the structures of VF_3 and MnF_3.

V. G. Solomonik, J. E. Boggs, and J. F. Stanton, *J. Phys. Chem. 103,* 838, 1999.
M. Hargittai, B. Réffy, M. Kolonits, C. J. Marsden, and J.-L. Heully, *J. Am. Chem. Soc. 119,* 9042, 1997.

These papers discuss the affect of core distortion on the geometry of transition metal molecules.

R. J. Gillespie, I. Bytheway, T.-H. Tang, and R. F. W. Bader, *Inorg. Chem. 35,* 3954, 1996.
R.J. Gillespie, D. Bayles, J. Platts, G. L. Heard, and R. F. W. Bader, *J. Phys. Chem. A. 102,* 3407, 1998.

Additional numerical data not given in the foregoing works can be found in the following articles, where further references are quoted.

R. J. Gillespie and E. A. Robinson, *Adv. Mol. Struct. Res. 4,* 1, 1998.
E. A. Robinson, S. A. Johnson, T.-H. Tang, and R. J. Gillespie, *Inorg. Chem. 36,* 3032, 1997.
E. A. Robinson, G. L. Heard, and R. J. Gillespie, *J. Mol. Struct. 485,* 305, 1999.

INDEX

ab initio calculations, 79–82,
actinide elements, 3
AIM theory, 154
   atomic properties and, 153–158
   bond properties and, 155–161
alkali metals, 2, 8–9
Allred, A. L., 15
ammonia, 19, 47, 173
angular wave function, 59–62
antibinding region, 135–136
antisymmetric wave function, 66, 80
atomic basin, 146
atomic charge, 153, 181
atomic dipole (moment), 45, 154–155
atomic interaction line, 151–153
atomic orbitals, 58–64
atomic properties, 153–159
atomic radius, 14–16
atomic units, 140
atomic volume, 181
aufbau principle, 69, 76

back-bonding, 38–39
Bartell, L.S., 116–118, 121
Bartlett, N., 21
basis functions, 81
Bent, Henry, 89–90, 122
bent bond model, 6, 100, 106
Berlin, 135–136
binding region, 135–136
Bohr, Neils, 6, 51–52
Bohr radius, 140
bond critical point, 147–151, 167, 183
bond dipole (moment), 46

bond dissociation energy, 39–42
bond enthalpy, 39–42
bond energy, 40
bond length, 27–30
   LCP model and, 119–120, 122–126
   polar bonds and, 37–38
   stereochemistry and, 84
bond line, 14
bond moment, 46
bond order, 30–33
bond path, 151–153, 183
bond strength, 25–27, 42–43, 183–184
   force constants and, 42–43
Bonham, 116
bonding charge concentration, 167
bonding radius, 157, 159, 161, 184
boranes (boron hydrides), 195–197
Born, 57–58, 80
Born-Oppenheimer approximation, 80
bosons, 66
boundary conditions, 55
building-up principle, 69
Butlerov, 3

charge transfer complexes, 19, 45, 155
chiral molecules, 6, 84
circles-on-a-sphere model, 89
cis isomer, 6
clamped nuclei, 80
coordinate analysis, 25
   force constants and, 42–43
coordinate bond, 19
coordination number, 35
   five, 243–250
   hypervalence and, 223–231

coordination number (*cont.*)
  LCP model and, 122–124
  seven or higher, 252–257
  six, 250–252
  up to four, 231–243
correlation energy, 80–81
Coulomb forces, 12–13
  Hellmann-Feynman theorem and, 134
Couper, 3
covalent bond, 10–13
Craddock, S., 28

dative bond, 19
Daudel, 139
Davisson, 53
de Broglie, 53, 57
delocalization energy, 32
density deformation (difference) functions,
    139–143
density functional theory (DFT), 81
diatomic hydrides, 157–161
dielectric constant, 44
dipole moment, 43–47
disphenoid, 108
domain model
  electron pairs and, 88–93
  localization and, 178–179
  electron pair domains and, 88–93, 93–99,
    106–110, 131
  orbital model comparison with, 106
donor-acceptor bonds, 19–20
double bond, 94, 204
double quartet model, 102–103, 171

Ebsworth, E. A. V., 28
eigenvalues, 149
Einstein, Albert, 50–51
electric dipole, 181
  moment of, 27, 43–47, 154–155, 161
electron density, 57–58
  atomic properties and, 151, 153–155
  bond properties and, 151–153, 155–161
  critical points and, 147–151,
  ellipticity and, 158–159
  Hellmann-Feynman theorem and, 134–136
  Laplacian of, 163–180
  probability and, 57–58
  representation of, 136–139
  topology of, 144–153

electron diffraction, 1, 27, 53, 84
electronegativity, 14–16
  bond angles and, 94, 96, 98–99
  bond length and, 37
  resonance and, 33
electron spin, 64–65, 86–87
electrostatic forces, 8–9
  electron pairs and, 86
ellipticity, 149, 157–159
enantiomeric forms, 6
ethene, 6, 99–100, 106, 117, 150
ethyne, 101

fermions, 66
five electron pair valence shells, 106–110
force constant, 42–43
formal charge, 17–18
Fourier transform, 143–144
free radicals, 12

Gaussian 2 (G2), 81
geminal ligands, 116
geometric isomers, 6
Gerlach, 64–65
Gillespie, R. J., 85, 119
Glidewell, C., 117–118, 121
Goudsmit, 64–65
gradient path, 145–147, 151–153
gradient vector field, 145–147
ground state, 54, 81, 167–168

Hamiltonian operator, 54–55, 58
Hargittai, I., 28, 118
harmonic oscillator, 27
Hartree-Fock method, 80–81
Heisenberg, W., 53
Hellmann-Feynman theorem, 134–136
Huygens, C., 50
hybrid orbitals, 72–76
hybridization, 72–76, 106, 117
hydrogen atomic orbitals 58–64
hypervalent molecules, hypervalence, 20–23,
    223–231
hypovalent molecules, hypovalence, 20–23

interatomic surface, 151
ionic model, 9–10
ionic radii, 33–37
isobutene, 116–117

isoelectronic atoms and molecules, 17, 214
isomers, 47, 84
isotopes, 3

Kauzman, 68
Kekulé, 3, 5, 31, 76
Kepert, D. W., 91
ketones, 117
Kimball, 89
kinetic energy, 54–55
Kossel, W., 7–8

lactic acid, 6
Langmuir, Irving, 10–11
lanthanide elements, 3
Laplacian of the electron density, 163–180
    ammonia and, 173
    CIF$_3$ and, 173–174
    covalent molecules and, 174–176
    ionic molecules and, 176–178
    methane and, 173
    VSEPR model and, 170–178
    water and, 172–173
lattice energy, 9
le Bel, 5, 84
Lennard-Jones, J. E., 66
Lewis model, 1, 10–13, 23
    acid base complex and, 19
    electron pairs and, 89, 163
    hypervalence and, 20–23, 223–231
    limitations of, 23, 204–205
    polyatomic ions and, 17–18
    resonance structures and, 30–32
ligand close-packing (LCP) model 119–132
    bond lengths and, 122–124
    bond angles and, 126–128
    coordination number and, 122–124
    lone pairs and, 126–128
    VESPR model comparison with, 132
ligand-lone pair coordination number (LLPCN),
    223
    five, 243–250
    hypervalence and, 223–231
    seven or higher, 252–257
    six, 250–252
    up to four, 231–243
linear combination of atomic orbitals (LCAO),
    71
line spectrum, 51

Linnett, J. W., 87, 102, 171
localization, 171, 178–179, 197, 218
lone pairs (nonbonding electron pairs), 11,
    94–98

Maxwell, 50
Mendeleev, D. I., 2–3
methane, 173
Mingos, D. M. P., 183
Moissan, 21
molecular graph, 152
molecular orbital (MO) method, 76–81
Moseley, 6
multiple bonds, 99–106
    resonance structures and, 30–33

natural bond orbital (NBO) method, 153–154
Newton, Isaac, 50, 54
noble gases, 2, 7
    ionic model and, 8
    Lewis model and, 10–11
    octet rule and, 21
nonbonding charge concentration, 167, 178
non-bonding electron pairs. *See lone pairs*
Nyholm, 85

octet rule, 11–12
    exceptions to, 20–23
    hypervalence and, 223–231
1,3 radii, 117–118, 121
optical isomers, 5, 84
orbitals, 58–64
    atomic, 58–64
    hybrid, 72–76
    hypervalence and, 223–231
    valence bond (VB) method and, 71–76
oxidation number, 18–19

particle-in-a-box model, 55–57
Pauling, Linus, 15, 21, 29
    back-bonding and, 38–39
    ionic radii and, 33, 36
    polar bonds and, 37
    resonance structures and, 31
Pauli exclusion principle, 67
Pauli force, 68, 113
Pauli principle, 49, 66–69
    electron distribution and, 85–88
    electron pairs and, 85–88, 163

Pauli principle (*cont.*)
  LCP model and, 113
  VSEPR model and, 85–88
period 2 molecules
  chlorides, 188–190
  fluorides, 184–188, 220–221
  hydrides, 190, 192–197
  hydroxo molecules, 198–202
  polar multiple bonds and, 202–209
period 3–6 molecules
  hypervalent, 223–231
  LLPCN of five, 243–250
  LLPCN of seven or higher, 252–257
  LLPCN of six, 250–252
  LLPCN up to four, 231–243
periodic table, 2–8, 67, 157–161
phosphazenes, 243
photons, 5–6, 50–51, 53, 57–58
Planck, Max, 50–51, 55
Platonic solids, 91
points-on-a-sphere model, 89, 91
polar bonds, 14–16, 44
  bond energy and, 37–38
  covalent radii and, 30
  hypervalence and, 223–231
  lengths of, 37–38
  multiple, 202–209
polyatomicions, 17–18, 25
potential energy, 13, 26, 56
  repulsion and, 121–122
Powell, 84–85
principal valence, 11
principle of maximum overlap, 71–73, 76–78,
    101
probability, 50–51, 57–58

quantization, 50–51
quantum mechanics, 1, 49, 82
  ab initio calculations and, 79–81
  atomic orbitals and, 58–64
  bonding models and, 71–79
  electron density and, 57–58, 88
  multielectron atoms and, 69–71
  Schrodinger wave equation and, 53–57
  spin and, 64–65
  wave function and, 53–58

radioactivity, 2
Ramsay, 21

rank (of a critical point), 149
Rankin, D. W. H., 28
resonance structures, 30–33
  back-bonding and, 38–39
Robinson, E. A., 119
Rochow, E. G., 15
rule of eight, 11. *See also octet rule*
rule of two, 11–12, 88
Rutherford, 6, 51

sawhorse (disphenoidal) geometry, 109
Schomaker, V., 29, 37
Schomaker-Stevenson equation, 37
Schrödinger equation, 49, 53–57
  hydrogen atomic orbitals and, 58
  probability density and, 57–58
  wave function and, 53–58
self-consistent field (SCF) method, 80
semiempirical calculations, 79
Shannon, 35–36
shell model, 6–8
Sidgwick, N. V., 84–85
Signature (of a critical point), 149
SI units, 28, 44, 140
Slater determinant, 80
sodium chloride, 8, 10, 35–36
spin (electron) 64, 65
  double-quartet model and, 102–103
  electron distribution and, 87
  electron pairs and, 88
  spin angular momentum, 64
  VSEPR model and, 85–88, 171
standard deformation density, 140–143
steric model, 116, 254
  see also ligand close-packing (LCP) model
Stern, 64–65
Stevenson, D. P., 29, 37
stretching force constant, 42–43
structural formulae, 3, 5
symmetric wave function, 66

tangent sphere model, 90, 93, 122
Thomson, 6
three center four electron (3c,4e) bond model,
    228–229, 245
topology of the electron density, 144
  atoms, 151
  bond paths, 151–153
  critical points, 147–151

gradient path, 145–147
gradient vector field, 145–147
molecular graph, 152
torsional angles, 25, 84
transition metals, 111, 257
triple bonds, 94, 104–105

Uhlenbeck, 64–65
Uncertainty principle, 53–54

valence, 1–2
hypervalent/hypovalent molecules and, 20–23,
223–231
ionic model and, 8–10
Lewis model and, 10–13. *See also* Lewis
model
oxidation number and, 18–19
polyatomic ions and, 17–18
shell model and, 7–8
structural formulae and, 3, 5
valence shell, 8
valence shell charge concentration (VSCC),
165–170
van der Waals radius, 113–116, 118, 121,
130–131
van't Hoff, J. H., 5, 84

vibrational energy, 25–26
bond length and, 27
force constants and, 42–43
VSEPR (valence shell electron pair repulsion)
model, 84–111
Laplacian and, 170–178
LCP model comparison with, 132
limitations of, 110–111
electron pairlocalization and, 178–179
pair domains and, 88–110

Wade, 183
wave function, 49–50, 82
bonding models and, 71–79
electron density and, 57–58
orbitals and, 58–64
Pauli principle and, 64, 66–69
Schrodinger equation and, 53–57
Werner, 84

X-ray crystallography, 1, 15, 25, 53, 82, 84
electron density and, 136, 143–144
ionic radii and, 35

Young, 50–51

zero-point energy, 57

FORMULA INDEX

Al$_2$Cl$_6$, 230, 251
AlCl$_3$, 22, 235–237, 240
AlCl$_4^-$, 241
AlF$_3$, 232–233
AlH, 160

As(CH$_3$)$_5$, 244
AsF$_5$, 108

B$_2$H$_6$, 195
B$_5$H$_9$, 197
B$_6$H$_6$, 197
BCl$_3$, 16, 18–19, 188–189, 191, 198, 240
BCl$_4^-$, 18
BF$_3$, 2, 22, 38, 121, 123, 176, 184, 186–188,
    190–192, 198
BF$_3$ · CO, 153
BF$_3$ · NH$_3$, 23
BF$_4^-$, 84, 93, 198
BH, 160
(BH$_2$)$_2$O, 217
BH$_3$, 74, 194–195
BH$_4^-$, 195
B(OH)$_3$, 198
B(OMe)$_3$, 200

Be$_4$O, 219
BeCl$_2$, 22, 188–189, 198
BeCl$_2$(OC$_2$H$_5$)$_2$, 23
BeF$_2$, 2, 14, 121, 176–177, 184, 186
BeH, 160
BeH$_2$, 74, 195
(BeH$_2$)$_2$O, 217

Bi(CH$_3$)$_5$, 244

BrF$_3$, 108
BrF$_4^-$, 97
BrF$_5$, 21, 255
BrF$_6^-$, 250, 254–255

C$_2$H$_2$, 174–175
C$_2$H$_4$, 174–175, 195
C$_2$H$_6$, 175
CH$_3$PF$_4$, 246

CCl$_3^+$, 203–204
CCl$_4$, 46, 188–190, 198
CF, 125–126
CF$_3^+$, 125–126, 198, 204, 246
CF$_3$OH, 205
(CF$_3$)PF$_2$, 246
(CF$_3$)$_n$PCl$_{5-n}$, 246
(CF$_3$)$_n$PF$_{5-n}$, 246
CF$_4$, 2, 11–12, 121, 176–177, 184–186, 188,
    190, 198, 205
CH, 101, 160
(CH$_3$)$_2$O, 96
(CH$_3$)$_2$C=CH$_2$, 117
(CH$_3$)$_2$O, 7, 129, 218
(CH$_3$)$_3$N, 96
(CH$_3$)$_3$P=CH$_2$, 244
(CH$_3$)$_4$N$^+$, 254
CH$_3$CH$_2$Cl, 16
(CH$_3$COO)$_6$, 220
CH$_3$Li, 244
(CH$_3$O)$_2$SO$_2$, 106
CH$_3$OH, 206

CH$_3$OTeF$_5$, 254
[(CH$_3$)$_3$Si]$_2$O, 218
CH$_3$SO$_2$CH$_3$, 118
CH$_4$, 40, 74, 87, 93, 173
ClF$_3$, 108
CO, 45–46, 125, 138–139, 147, 202–209
CO$_3^-$, 205–28
C(OMe)$_4$, 200
C(OPh)$_4$, 200

CaF$_2$, 9

Cl$_2$, 15, 20, 37, 138, 147
Cl$_2$CO, 204
Cl$_2$O, 129, 217
ClF$_3$, 178
ClF$_4^-$, 97
ClF$_6^+$, 250
ClH, 160
ClH$_2$C−CH$_3$, 16
Cl$_3^-$, 115
ClSO$_2$C$_6$H$_5$, 118
ClSO$_2$CCl$_3$, 118
ClSO$_2$CF$_3$, 118
ClSO$_2$Cl, 118

CrCl$_2$, 111

F$_2$, 2, 11–12, 185
F$_2$CO, 125, 204
F$_2$O, 216–217
F$_3$B · NH$_3$, 17
F$_3$CO$^-$, 207
F$_3$PH$_2$, 246–247
F$_3$PO, 125
F$_4$PH, 246
FCl, 189–190
FH$_2$CO, 207
FSO$_2$CH$_3$, 118
FSO$_2$F, 118
FSO$_2$OCH$_3$, 118

(GeH$_3$)$_2$O, 218

H$_2$, 15, 37, 216
H$_2$CO, 202

H$_2$CSF$_4$, 248
H$_2$N−NH$_2$, 16
H$_2$O, 73, 128, 172–173, 178
H$_2$S, 236
H$_2$SO$_4$, 22
H$_3$O$^+$, 219
H$_3$C−CH$_3$, 16
H$_3$CO$^-$, 207
H$_3$N, 19
H$_3$O$^+$, 93
H$_3$PO$_4$, 22
(H$_3$Si)$_2$O, 216, 218
H$_5$IO$_6$, 21
HCl, 6, 37, 45, 236
HClC=CHCl, 6
HF, 87, 101
HN=SF$_4$, 249
(HO)$_2$O, 217
HSCH$_3$, 29

I$_2$Cl$_6$, 251
ICl$_2^-$, 108
ICl$_4^-$, 97
IF$_4^-$, 97, 108
IF$_6^+$, 250, 254, 257
IF$_7$, 21, 253
IF$_8^-$, 256
IOF$_6^-$, 253

KBr, 34
KI, 34

KrF$_2$, 21

Li, 62, 184–197
Li$_2$O, 110, 128–129, 217–218
LiCl, 188–190, 198
LiF, 2, 14, 22, 176, 184–187
LiH, 157, 159–160
LiOH, 128

MgCl$_2$, 8–9
MgH, 160

MnCl$_2$, 111

$NaBF_4$, 17
$NaBr$, 34
$NaCl$, 8, 34, 36
NaF, 235
NaH, 157, 159–160
NaI, 34

$N(CF_3)_3$, 212
$N(CH_3)_3$, 110, 218
$N(CH_3)_4^+$, 213–214
NCl_3, 93, 189, 210–211
NF_3, 2, 11–12, 47, 99, 185, 210–211, 218, 238
NF_4^+, 213
NH, 160
$(NH_2)_2O$, 217
NH_3, 47, 93, 173, 178, 192, 211, 218
NH_4^+, 84, 93
NH_4Cl, 17
$N(SCF_3)_3$, 212
$N(SiH_3)_3$, 216, 218

OCF_3^-, 204–205
$O(CH_3)_2$, 110
OCl_2, 96, 110, 189
OF_2, 2, 11–12, 185
OH, 160, 250
ONF_3, 214, 225
OSF_4, 248

$P(C_6H_5)_5$, 223
$P(CH_3)_5$, 243–244
PCl_3, 93, 99, 235–237
PCl_4^+, 124, 241
PCl_5, 20, 108, 209, 224
PCl_6^-, 124, 209, 250
PF_3, 99, 130, 177, 232–233, 238
PF_4^+, 108, 125, 130, 241
PF_4H, 247
PF_5, 22, 108, 209, 244–245
PF_6^-, 209, 250
PH, 160
PI_3, 99
$POCl_3$, 16
POF_3, 130

PtF_6, 21

RbI, 34

$S(CH_3)_2$, 29
SCl_2, 136–138, 151, 167, 169, 178, 235–237
SF_2, 232–233
SF_4, 20, 108, 249
SF_5CF_3, 250
SF_5Cl, 250
SF_5NF_2, 250
SF_6, 20, 22, 46, 224, 230, 250
SH, 160
SO_2, 22
$SO_2(OH)_2$, 225
SO_3, 19, 22

$Sb_2F_{11}^-$, 220–221
$Sb(C_6H_5)_5$, 244
$SbCl_3$, 99
$SbCl_5$, 108
SbF_3, 99
SbF_6^-, 250
$Sb(OH)_6^-$, 250

$ScCl_2$, 111

$SeCl_2$, 255
$SeCl_6^{2-}$, 257
SeF_4, 108
SeF_6, 250, 254, 257

$SiCl_4$, 93, 235–237
SiF_4, 39, 124, 232–233, 235
SiF_6^{2-}, 124
SiH_3^+, 129, 217–218
SiO_2, 241
$Si(OH)_4$, 241

$Te(CH_3)_2Cl_2$, 223
TeF_6, 250
TeF_7^-, 253
$TeOF_6$, 253
$Te(OH)_6$, 250
$Te(OMe)_2F_5^-$, 254
$Te(OMe)F_6^-$, 253–254

$TiCl_2$, 111

VCl$_2$, 111

XeF$_2$, 108
XeF$_3$$^+$, 108
XeF$_4$, 21, 97

XeF$_6$, 21, 225, 254–255, 257
XeF$_8$$^{2-}$, 256–257
XeO$_3$, 21
XeOF$_4$, 21, 252
XePtF$_6$, 21